The Story of Creation as Told by Theology and by Science

T. S. Ackland

THE STORY OF CREATION AS TOLD BY THEOLOGY AND BY
SCIENCE.

BY T. S. ACKLAND, M.A.,

FORMERLY FELLOW OF CLARE COLLEGE, CAMBRIDGE; VICAR
OF WOLD NEWTON, YORKSHIRE.

"SIRS, YE ARE BRETHREN: WHY DO YE WRONG ONE TO
ANOTHER? "

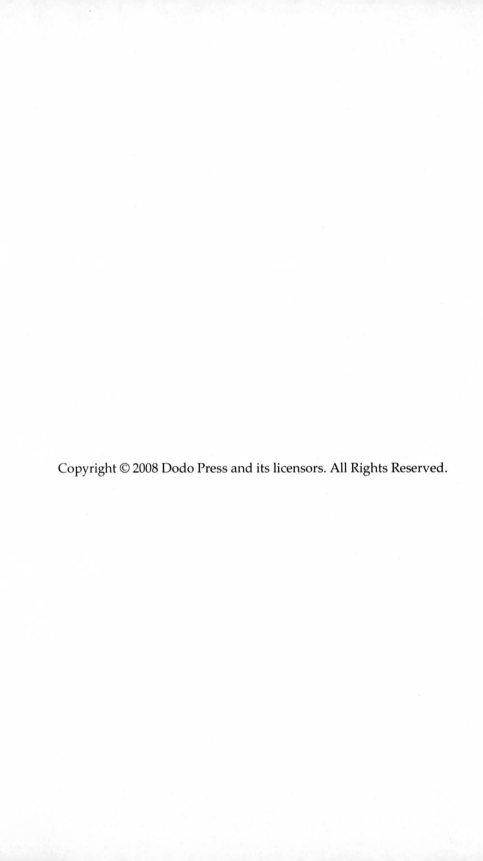

CONTENTS.

CHAPTER I.

THE CASE STATED.

The History of the Creation with which the Bible commences, is not a mere incidental appendage to God's Revelation, but constitutes the foundation on which the whole of that Revelation is based. Setting forth as it does the relation in which man stands to God as his Maker, and to the world which God formed for his abode, it forms a necessary introduction to all that God has seen fit to reveal to us with reference to His dispensations of Providence and of Grace.

It is, however, not uncommonly asserted that this history cannot be reconciled with a vast number of facts which modern science has revealed to us, and with theories based on observed facts, and recommended by the unquestioned ability of the men by whom they have been brought forward. At first sight there does seem to be some ground for this assertion. Geology, for instance, makes us acquainted with strata of rock of various kinds, arranged in exact order, and of an aggregate thickness of many miles, which are filled with the remains of a wonderful series of plants and animals, these remains not being promiscuously collected, but arranged in an unvarying order. It seems impossible that all these plants and animals could have lived and died, and been imbedded in the rocks in this exact succession, in six of our ordinary days. Astronomy directs our attention to changes now going on in the starry heavens which occupy ages in their development, and points to traces in the constitution of our own world which seem to indicate that it was formed by analogous means. Physiology reveals to us the fact that the different varieties of plants and animals now in existence are not separated from each other by well defined lines of demarcation, but shade into each other by almost imperceptible gradations; and geological researches show that while the existing species of animals are the representatives of those which lived and died at a period in which we can find no traces of man, they are not identical with them, but that either the old species must have died out, and been replaced by a fresh creation, or a considerable change must have taken place in the course of ages. These facts are held to be incompatible with the account of creation given by Moses, and hence it is inferred that a record, which appears to be so widely at variance with admitted facts, cannot be entitled to the authority which is claimed for it, as a fundamental portion of a Revelation made by the Creator Himself.

This difficulty is sometimes met by the assertion that the Bible was not given to us to teach us Science, but to convey to us certain information which was essential to our moral welfare, and which we could not obtain by any other means; that these discrepancies do not in any way interfere with that portion of those truths which is involved in the History of Creation, but that, however the narrative may be viewed as far as regards its details, the facts that God is the Creator of all things visible and invisible, that He is a Being of infinite Wisdom, Power, and Love, and that He has placed man in a peculiar relation to Himself, remain unaffected. On this ground it is often urged that we may pass over scientific inaccuracies as matters of no great importance.

Theologians are by no means agreed as to the nature and limits of that inspiration by which Holy Scripture was written. There are many who think that in matters purely incidental to its main object, and lying within the reach of human faculties, the sacred writers were left to the ordinary sources of information, and that many alleged difficulties may be removed by this view.

But whatever may be thought of the application of this hypothesis to some parts of the Bible, there are others to which it is plainly inapplicable, and of these the narrative of the Creation is evidently one. No theory of limited inspiration can be admitted to explain any supposed inaccuracies in that narrative. It cannot be liable to those imperfections which are inevitable when men have to obtain knowledge by the ordinary means, because there were no ordinary means by which such information could be obtained. The most carefully preserved records, the oldest traditions could not extend backwards beyond the moment when the first man awoke to conscious existence. For every thing beyond that point the only source of knowledge available was information derived from the Creator Himself. It may be that a revelation of this character was made to Adam in the days of his innocence, that it was carefully handed down to his descendants, and that Moses, under the divine direction, incorporated it into his history; or it may have been directly communicated to Moses by special inspiration—that matters not—but a divine revelation it must have been, or it is nothing; the dream of a poet, or the theory of a philosopher, if we can believe that such a philosopher existed at such a time. But if it be indeed a revelation from the Creator Himself, we cannot imagine that He could fall into any error, or sanction any misrepresentation with reference even to the smallest detail of His own work.

If then there are really any errors in this record—any assertions which the discoveries of science have proved to be untrue, we cannot account for them on any theory of limited inspiration. A single proved error would be fatal to the authority of the whole narrative. But, on the other hand, we are not justified in expecting such an account of the Creation as would commend itself to the scientific intellect of the present day. When we attempt to form a judgment upon it. We must look not only to its alleged author, but also to the purposes for which, the circumstances under which, and the persons to whom it was given. In these we may expect to meet with many limitations. It was not designed for the communication of scientific knowledge, it was necessarily conveyed in human language, and addressed to human intelligence, that language and that intelligence being, not as they are now, but as they were, taking the latest possible date that can be assigned to it, considerably more than three thousand years ago.

This last consideration affects not only the record itself, but also our facilities for understanding and forming a judgment upon it. We have to contend with difficulties of interpretation arising from our inability fully to realize the circumstances under which it was given, and to place ourselves in the mental position of its original recipients. Owing to our want of this power it may well happen, that though we are in possession of vastly increased knowledge, we may be far more liable to fall into error in some directions, in the interpretation of it, than those to whom it was originally addressed.

An additional difficulty arises from the circumstance that our knowledge, wonderfully as it has been increased of late, is yet very far from complete, and is probably in many cases still mixed with error. Hence it may very well happen that where there is complete harmony between the history and the facts, we may suspect discord owing to our misunderstanding of the record, or our misconception of the facts. In order that the harmony may be recognized in its fulness, there must be a perfect understanding of the record, and a perfect knowledge of the facts. But from both of these we are probably at present very far removed.

If a person who was a thorough master of some science undertook to write a treatise for the purpose of teaching children the rudiments of that science, we should expect, and the more strongly if the author were a master of language as well as of science, that his work should contain indications of a master's hand. We should expect that while the book conveyed clearly and simply to the minds of those for whom it was written, the truths which it was intended to teach, it should also convey to the more educated reader some intimations of a deeper knowledge on the part of its author. The choice of a word, the turn of a phrase, the order in which facts were arranged, the occurrence here and there of a sentence which an ordinary reader would pass over as unimportant, would to such a person be indications of trains of thought far more profound than those which appeared on the surface. And this recognition would be proportional to two things—the amount of scientific knowledge possessed by the reader, and his mastery of the language in which the book was written.

Such, then, are the characteristics which we may expect to find in the Record of Creation, if it be indeed, as we believe, a revelation from God, made to men in a very low stage of intellectual development. In order that we may be able to form a satisfactory judgment of it, it will be well for us to consider a little in detail two classes of difficulties. 1. Those which belong to the Revelation itself, arising from the limitations to which it was necessarily subject in its delivery. 2. Those which arise from our imperfect knowledge of the language in which it is written, and from our inability to place ourselves in the intellectual position of those to whom it was originally given.

1. When this record was committed to writing, language was in a very different condition from that in which it is now. We have an account of the first recorded exercise of the faculty of speech in Gen. ii. 19. Adam first used it to give names to all the living creatures as they passed in review before him. In accordance with this statement it appears, from the researches of philologists, that language in its earliest state was entirely, or almost entirely limited to words denoting sensible objects and actions. It seems probable that these names were derived from radicals expressing general ideas [Footnote: Max Muller's Lectures on the Science of Language, First Series Lect. viii. ix.]; but there is reason to doubt whether these radicals ever had a formal existence as words—they seem rather to have been the mental stock out of which words were produced. But the human mind had from the first powers for the exercise of which this limited vocabulary was insufficient. Even in the outer world

4

there was much which was the object of reason and inference rather than of sense, while the whole world of consciousness was entirely unprovided with the means of expression. To meet this difficulty words, which originally denoted objects of sense, were used figuratively to express ideas which bore some resemblance or analogy, real or fancied, to their original significance. As time passed on this difficulty was gradually diminished: synonyms crept into all languages from various sources, and when once adopted, they were in many cases gradually differentiated, the various senses which the original word had borne were portioned off among them, and increased precision was thus obtained.

But in the infancy of mankind the figurative system was in full operation. Hence, all early documents have a strong tinge of the poetic element. Poetry, strictly so called, probably had not as yet a separate existence; but the whole spoken and written language was permeated by that poetic spirit which delights in tracing subtle analogies, and in expressing the invisible by means of the visible. The translation of the Sanscrit Hymns, which has recently appeared [Footnote: Hymns of the Big Veda Sanhita, translated by Max Muller, vol. i.], furnishes a most valuable illustration of this state of thought and of language. These hymns are probably nearly coeval with the Pentateuch. They were the production of a different branch of the human family, and indicate a different tone of thought, but they bring out very clearly the figurative character of primitive language, abounding in fanciful descriptions of natural phenomena, which, when their metaphorical, character was forgotten, passed by an easy transition into the graceful myths and legends of early Greece.

Then there was a poverty in these primitive vocabularies even in reference to sensible objects, which in many cases rendered it necessary to employ the same word in more or less extensive significations, and in the Semitic languages the power of inflexion was in some directions very limited. This limitation is most remarkable in the forms used for the expression of time. One form alone was available to express those modifications which are indicated by the imperfect, perfect, pluperfect, and aorist tenses of the classical languages.

Instances of all these sources of uncertainty meet us very early in Genesis. In the very first verse we have a word, [Hebrew script], which has great latitude of meaning. It is either the earth as a whole (ver. 1), or the land as distinguished from the water (ver. 10), or a particular country (ii. 11). In many cases, as in all these, the context at once determines the sense to be chosen; but there are other cases in which considerable difficulty arises. The whole question of the universality of the deluge turns, in a great degree, upon the signification which is assigned to this same word in the sixth and following chapters. In the second verse we have another word, [Hebrew script], which is capable of various interpretations. It is used throughout the Bible in the three distinct meanings of "wind, " "breath, " and "spirit. " Where we read, "And the Spirit of God moved upon the face of the waters, " the Jewish paraphrase is, "And a wind of God (i. e. a great wind) moved, " &c. Here there is nothing in the context to assist us in determining the sense to be chosen; but, as will be seen in the sequel, modern science indicates that the Jewish interpretation is untenable, and that our translation is, consequently, the correct one. As an instance of confusion of time, we may refer to ii. 19. In our translation this verse seems to place the creation of animals after that of man; but in xii. 1, the very same form is translated by the pluperfect, "Now the Lord had said unto Abram. " It ought evidently to be translated in the same way here: "And out of the ground the Lord God had formed, " &c. In ii. 5, on the other hand, the pluperfect might with advantage have given place to another form: "For the Lord God did not cause it to rain. " The phenomenon referred to appears to have been local and temporary. Had the pluperfect been omitted in one case and supplied in the other two sources of apparent difficulty would have been removed.

It is very clear, then, that there could be no approach to scientific accuracy in a narrative written in such a language as this. Such accuracy is, in fact, attainable only in proportion, as science has moulded language for its own purposes. But language is at all times an index of the general mental condition of the people who use it, and so the knowledge and the ideas of the men of these primitive times must have been extremely limited in all those directions with which we have to do. Accordingly, we find no trace of any doubt whether the information with reference to external objects which was received through the senses was in all cases to be depended on. There can be little doubt that to those early observers the sky was a solid vault, on the face of which the sun, moon, and planets moved in their appointed courses; the stars were points of light, golden studs in the azure canopy; the sun and moon were just as large as they appeared to be, and the earth was a solid immovable plane of comparatively small extent. At the time of the Exodus, it seems clear that, even among a people so far advanced as the Egyptians, all that lay beyond the mountains which bounded their land on the west was believed to belong not to living men, but to disembodied spirits. It was the terrible country through which the souls of the departed made their arduous way to the Hall of Judgment [Footnote: "The Nations Around, " pp. 49, 50.] Accordingly, we find that the Egyptians made no attempt to extend the limits of their empire in this direction, while the monarchs of the Mesopotamian region seem to have been equally unambitious of conquest beyond the mountain ranges which bounded the valley of the Tigris on the east. Mesopotamia, then, on the east, Egypt on the west, Armenia and Asia Minor on the north, and Arabia on the south, seem, in the view of the contemporaries of Moses, to have been the utmost regions of the world. Ignorant as they were of any countries beyond these, they were, of course, equally ignorant of the numberless varieties of plants and animals that were to be found in them, and with which we are familiar. Mining was not unknown, but the mines were few and superficial; they could not reveal much of the structure of the earth, and what little they did reveal passed unnoticed. Nothing was known of the successive beds of rock which form the crust of the earth, of the fossils with which they abound, or of the gradual changes to Which they are still subject. If any one had told the men of that generation that the solid earth on which they stood, or the everlasting hills which surrounded them, were undergoing slow but steady modifications, he would have been looked upon as a madman.

A revelation, then, addressed to men whose language, whose intellectual powers, and whose stock of ideas were thus limited, must of itself also necessarily have been both limited and destitute of precision. It could only deal with things with which they had some acquaintance, or of which they could form some idea, while, from the character of the language, and the extreme brevity of the record, the treatment of even these few subjects must have been of a vague and indefinite character. Traces of a deeper knowledge there might be, but they would not lie upon the surface. They must be carefully sought for, and then they would be discernible only by those who were in possession of the key which would unlock their hidden secrets.

Such are the limitations under which the revelation was necessarily given. We have now to consider our own especial difficulties, the obstacles which stand in our way when we would discover for ourselves all the information which the record is capable of conveying. For if this record be, as we believe, the work of the Great Architect of the Universe, then it is probable that its every detail is significant; that wherever it was possible words were chosen which, when scrutinized, would convey much more information than appeared on the surface. The great problem for us to solve is, What are the difficulties which stand in our way when we would seek this knowledge, and what are the means by which those difficulties may be surmounted, and the hidden treasure displayed?

Our first difficulty arises from a matter which, viewed in another light, is one of our greatest blessings. We are familiar with the Record through the medium of our own noble version. Probably it is impossible for any translation more exactly to represent the original as it presented itself in the first instance to the minds of those to whom it was addressed. Accordingly we learn it in our earliest childhood; its majestic phrases imprint themselves on our memory; our undeveloped minds seem capable of taking in all that it was intended to convey, and so the impressions formed of it in our infancy abide with us all our days. We are contented with them, and do not trouble ourselves to inquire whether there is not something beyond, which we have not realized.

All this time we forget that, excellent as it is, it is after all only a translation, and that the very best translation cannot represent in their fulness the ideas embodied in the original. Etymological relations between words often give a force and meaning to a sentence which it is impossible to transfuse into another language, because the same relations do not exist between the words which we are constrained to employ. Then there is an intimate relation between men's thoughts and the language which they habitually use, so that those thoughts cannot be perfectly expressed in a language whose character is different. Again in every language there are many words which bear several cognate senses, which may be represented by as many different words in the language of the translation; so that if the best word is chosen, much of the fulness of the original must be lost; while it may so happen that the selected word has also a variety of significations, which do not correspond with the varying meanings of the original word, and thus senses may be ascribed to the original which it will not bear, because the reader annexes to the word in the translation a sense different from that in which it corresponds to the original word. To all these sources of imperfection must be added the fact that our translation was made at a time when science was not yet sufficiently developed to exercise any influence upon it. There was nothing to induce the translators to attempt, where it was possible, to preserve any indications of a deeper meaning, because they had no reason to suspect that any such deeper meaning existed, or that any indications of such a meaning were to be found.

To the difficulties of translation must be added the difficulties of accumulated tradition. The characteristics which mark our own childish intellect are apparent also in the collective intellect of the human race in its earlier and ruder development. There are two characteristics of the human mind in this condition, which have had a very great effect on the interpretation of this portion of the Bible.

The first of these is the impatience of doubt and uncertainty. The power of recognizing the imperfection of our knowledge, and the consequent necessity of suspending our judgment, is a power which is only gradually acquired with the accumulation of experience. The young untrained mind finds it difficult to realize the truth that any information communicated to it is not altogether within the grasp of its faculties. It must attach some definite meaning to the words; it must image to itself some way in which great events were brought about, great works were accomplished. It finds it difficult to realize a fact as accomplished, unless it can also picture to itself some way in which it might have been effected. For this purpose such knowledge as it has at its command is employed, and where that fails recourse is had to the imagination to supply the deficiency. Thus it has been with ourselves in our childhood, and thus it was in the childhood of the world. Knowledge was indeed sought, but it was not sought in the right way, and so the search often resulted in error, and this error produced its effect in the interpretation of the passage in question. The old school of inquirers started from certain abstract principles, and endeavoared to reduce the results of observation to conformity with those principles. This was the case with astronomy. The old astronomers taking as axioms the two assumptions that everything connected with the heavenly bodies must be perfect, and that the circle is the only perfect figure, easily satisfied themselves that the orbits of all the heavenly bodies must be circles. Hence came the

"Cycle on epicycle, orb on orb, "

by which they sought to account for the phenomena which they observed. When once the method was changed, when once it had occurred to Kepler that, as it seemed to be impossible to account for

9

the apparent motion of Mars by any theory of circular orbits, it might be worth while to try to ascertain by observation what its orbit really was, a few years of patient labour sufficed to solve the problem.

It was science such as this, then, that our forefathers brought to the interpretation of the Mosaic Record, and the consequence was that when, from time to time, facts were casually brought to light which might have led the way to vast discoveries, their true significance was never discerned; all that was sought from them was some additional support to the old views. Thus sometimes gigantic bones were exhumed: without investigation, it was at once assumed that they were human bones, and they were brought forward to prove the truth of the statement, "There were also giants in the earth in those days. " Sea-shells were found on mountain sides, far from and high above the sea—they were evidences of the Deluge.

The second characteristic of that state of mind is its admiration of the startling and the vast. In these alone it recognizes the tokens of unlimited power. It is unable to appreciate those more majestic manifestations of power which are discerned by the enlightened eye, when a stupendous scheme is developed, gradually and imperceptibly, but without pause or hesitation through a long succession of ages; when a multitude of seemingly discordant elements are at last brought together in a perfect work; when a power, unseen and unnoticed, slowly but surely overrules the working of ten thousand apparently independent agents, through a thousand generations, and moulds their separate works into one harmonious whole. Such a manifestation of power as this was beyond the grasp of the untrained mind; but to such intellects there was something irresistibly fascinating in the idea of a world rising into perfect existence in a moment, of innumerable hosts of living creatures called into being at a word. Such was the meaning of the account of creation which naturally suggested itself to the untrained mind, and there was nothing in science in those early days to throw any doubt upon it, and so this belief was unhesitatingly and almost universally adopted. Here and there, indeed, some man of deeper thought than his brethren, such as St. Augustine [Footnote: See St. Augustine, "De Genesi ad Literam, " Liber Imperfectus, and Libri Duodecim, and also "Confessionum" Liber xiii.], suspected that there might be more in that seemingly simple record than was generally acknowledged; but such men had no means of verifying their conjectures, and their number was very small. For three thousand years the old view was practically unquestioned, it received the tacit sanction of the Church, it gradually became identified in the minds of all with the record itself, and was as much an article of faith as the very Creed.

This was the state of things, when at last science awoke from its long slumber, and began for the first time to employ its energies in the right direction. Very soon discoveries were made which startled the minds of all believers in the Bible. The first shock which the old belief sustained was from the establishment of the Copernican view of the Solar System. That the world was the immovable centre of the universe, around which sun, moon, and planets moved in their appointed courses, was universally held to be the express teaching of the Bible; and when Galileo ventured to maintain the new views in Italy, the Roman Curia took up the question, and by the agency of the Inquisition wrung from him a reluctant retractation of his so-called heresy. But it was of no avail. The new doctrine was true, and it could not be crushed. Fresh evidence of its truth was continually coming forward, till at last it was universally received. Then the defenders of the Bible had recourse to the suggestion that as the Bible was not intended to teach us science, such errors were of no consequence, But this argument, though perfectly sound with reference to such passages as Joshua x. 12-14, where an event is described as it appeared to those who witnessed it, is not admissible in such a passage as Psalm xcvi. 10, where the supposed immobility of the earth is alleged as a proof of God's sovereignty, and is made the foundation of the duty of proclaiming that sovereignty among the heathen. When the supposed proof was found to be a fallacy, the statement in support of which it was alleged would be more or less shaken. In such a passage, then, the theory of limited inspiration is evidently untenable. At last the only sensible course was adopted. Recourse was had to the original, and it was at once apparent that the supposed difficulty had no real existence, but that there was a very trifling inaccuracy in the translation; for that the word translated "shall not be moved" really signified "shall not be shaken or totter. " The same word is used in Psalm xvii. 5, "Hold up my goings in Thy paths, that my footsteps SLIP NOT. " Instead, then, of an error, we have an exact description of the earth's motion—a motion so steady and equable, that for thousands of years no single individual out of the myriads who were continually carried along by it had ever suspected its existence.

Well had it been for all if the lesson thus taught had been deeply laid to heart. But unhappily it was entirely unnoticed. Science pursued its way with increasing energy, and more facts were year by year brought to light which seemed entirely to contradict the teaching of the Bible, and again alarm and distrust sprung up in the minds of what, for want of a better name, we may perhaps be allowed to designate as the "Theological Party. " The power of the Church of Rome was by this time so far curtailed that the old means of repression were no longer available; but the old spirit survived, and not in Rome only. There was the same blind distrust, the same mistaken zeal for supposed truth, the same indignation which naturally arises when things which we hold precious are attacked, and, as it seems to us, without any sufficient reason.

There was indeed much to account for and even to justify the feelings of anger and alarm which were excited, for the time when these discoveries began to be brought prominently forward was the latter half of the last century. At that time the famous French Academy was doing its deadly work, and the new discoveries were gladly hailed by the infidel philosophers of France, as weapons against the Bible. But the reception given to these discoveries by the theological party, though partially justified by the circumstances of the times, was nevertheless very mischievous in its results. For though the new discoveries were hailed enthusiastically by the infidel school, a very large portion of the men by whom they were made, and of those who were convinced of their truth, were men of a very different character. They were simple earnest seekers after truth as it is displayed in God's works. Their belief in the Bible rested in most cases on the authority of others. They had not investigated for themselves its external evidences; in many cases they had neither the ability nor the opportunity to do so; nor had many of them as yet become practically familiar with that internal evidence which the faithful Christian carries within him, though in time they might have become so, had they not been driven into infidelity by the reception which was given to their discoveries. When men of this character were informed by those to whom they were accustomed to look up as teachers in religious matters, that the discoveries, of the truth of which they were so firmly convinced, and in which they took such justifiable pride, were contradictory to the teaching of the Bible, they were placed in a position of extreme difficulty. For this statement was, in fact, a demand made upon them that they should give up these discoveries as erroneous, or else renounce their belief in the Bible. But their belief in the Bible rested in the main on the authority of others; they felt themselves incompetent judges of the evidence on which it rested, while they were fully acquainted with, and

12

competent judges of, the grounds on which their own discoveries were based. The evidence on which they acted was, to their minds, quite as convincing as the Biblical evidence was to the minds of their antagonists. Two things, then, were pronounced incompatible by what seemed to be a competent authority; they could not adhere to both, and the natural consequence was that their assent was given to those statements which rested on evidence which they thoroughly understood, and the Bible was rejected. Thus it has come to pass that many of our scientific men, if not professed unbelievers, have yet learnt to look upon the Bible with suspicion and distrust. To some of them, as is evident from their writings, their position is a matter of profound sorrow.

There have, indeed, been many noble exceptions to this state of things. Many men whose pre-eminence in scientific knowledge and research is admitted by all, have yet clung in childlike trust to the Bible. They have recognized its authority, they have been satisfied that God's Word could not be in opposition to His Work, and they have been content to wait in unquestioning faith for the day when all that now seems dark and perplexing shall be made clear. But there have also been very many with whom this has not been the case, and their unbelief has not affected themselves alone. The knowledge of it has had a deadly effect upon thousands who were utterly incompetent to form any judgment on either theological or scientific subjects, but who gladly welcomed anything which would help to justify them to their own consciences in their refusal to submit themselves to a law which, in their ignorance, they deemed to be harsh and intolerable. There has also been another class of sufferers. Many persons who loved the Bible, but whose education, and, consequently, whose powers of judgment in the matter were very limited, have received very great injury from the doubt which has been thrown on its authority. Unable of themselves to form a judgment on the subject, they could not be unmoved by the opinion expressed by those whom they regarded as better informed than themselves. Hence their faith has received a shock always painful and dangerous, often perhaps fatal.

Many attempts have been made to overcome the difficulty which has thus arisen. When geologists first began to study the lessons which are to be learnt from fossils, a suggestion was made which, though it was soon shown to be untenable, has still perhaps a few supporters. It was said that these fossils were not what they seemed to be, the remains of creatures which once lived, but simple stones, fashioned from the first in their present form by the will of the Creator. But such an idea is at variance with all that either Nature or Revelation teaches us concerning God. All those who have any familiarity with the subject cannot but feel that the suggestion of such a solution of the difficulty is little short of a suggestion that the Almighty has stamped a lie upon the face of His own Work.

Another proposed solution, which for a time seemed satisfactory, assumed several successive creations and destructions of the world to have taken place in the interval between the first and second verses of Genesis. To these all the fossil remains were ascribed, while the present state of things was supposed to be the result of the operations recorded in the remainder of the chapter. But as geological knowledge advanced, it soon became clear that there were no breaks in the chain of life; no points at which one set of creatures had died out, while another had not yet arisen to fill up the void, but that all change had been gradual and progressive, and that species still living on the earth are identical with some which were in existence when the lowest tertiary strata were in process of formation—a time which must have been many thousand years prior to the appearance of man.

Other attempts have been made upon literary grounds. Hugh Miller [Footnote: Testimony of the Rocks.] carefully worked out a suggestion derived from a German source, that the history of Creation was presented to Moses in a series of six visions, which appeared to him as so many days with intervening nights. More recently Dr. Rorison [Footnote: In Answers to "Essays and Reviews. "] has maintained that the first chapter of Genesis is not a history at all, but a poem—"the Hymn of Creation. " There is, however, nothing in the chapter itself to confirm either of these views. When visions are recorded elsewhere we are told that they are visions, but no such hint is given us here. Nor do we find in the passage any of the characteristics of Hebrew poetry. It is inserted in an Historical document, and in the absence of any proof to the contrary, it is plainly itself also to be regarded as History.

But there remains yet one method to be attempted. If there is reason to believe that the Bible is the Word of God, just as the universe is His Work, then we may well expect that each of them will throw light upon and help us to a right understanding of the other. And if there be one part beyond all others in which this may be confidently looked for, it is that part in which the Divine Architect describes His own work. We know how difficult it is to understand a complicated process, or a complex piece of machinery, from a mere written description; and how our difficulty is lessened if we have the opportunity of inspecting the machinery or the process. Just in the same way we may expect to encounter difficulties, and to form erroneous conclusions when we study by itself such a document as the history of Creation, and we may well expect that those difficulties will be diminished, and those errors corrected by an examination of that material universe, the production of which it describes. And, on the other hand, if science—the study of the universe—is found to throw light upon and to receive light from the Bible, this is a fresh proof that the Bible and the universe are from the same source; the authority of the Bible is more firmly established, and the conclusions arrived at by men of science are confirmed.

But before this can be done to any good purpose, something is required from both the contending parties. The theological party must be prepared to sacrifice many an old opinion, many a cherished belief. Great care must be taken to discriminate between the genuine statements of the Mosaic Record, and the old interpretations which have been incorporated into and identified with those statements. Some, perhaps, may fear lest, in rejecting those interpretations, they may be setting at nought an authority to which they ought to submit, since these interpretations seem to have the sanction of the Church. But it can hardly be maintained that those promises of Divine guidance and protection from error which were given to the Church extended to such matters as this. No question of faith or duty is involved in the interpretation which we may give to the details of Creation. If there are some parts of the Bible in which the earliest interpretation is unquestionably the true one, there are also other parts, such as many of the prophecies, which became intelligible only when light was thrown upon them by subsequent events. And so it seems to be with the Record of Creation: it can only be rightly understood in proportion as we become acquainted with the details of the matters to which it refers. Any interpretation which was put upon it before those details were brought to light must of necessity be liable to error.

But something is also required of the opposite party. At the very threshold of the investigation they must be asked to lay aside, so far as is possible, those prejudices against the Bible which have naturally arisen in their minds from the obstinacy with which views, which they knew to be untenable, have been forced upon their acceptance as the undoubted teaching of God, so that they may enter upon the investigation with unbiassed minds. Then they must be careful to distinguish between established facts, and theories however probable. There is something very fascinating in a well constructed theory. Theories have again and again done such good service in opening the way, first, to the discovery, and then to the arrangement of facts, that we are very apt to assign to them an authority far beyond that to which they are really entitled. When, for instance, we have ascertained that a certain number of facts are explained by some particular theory, we are apt to assume prematurely, that the same theory must account for and be in harmony with all similar and related facts; or, if we have satisfied ourselves that certain results MAY have been produced in a particular way, we are in great danger of being led to conclude that they MUST have happened in that way. No mere theory can have any weight against a statement resting on solid evidence, but where the evidence is weak, or, what is practically the same thing, where the knowledge of that evidence is defective, a probable theory must carry great weight in influencing our judgment. Care must therefore be taken to keep theories in their proper place. Where we have to deal with well-established facts, any interpretations to which those facts may lead us may be taken as also established, but interpretations which are suggested by theories only must be regarded as provisional, and liable to future modification or rejection, as our knowledge increases.

The Mosaic Record itself, when carefully examined, seems to be peculiarly open to the process suggested. No doubt there is yet much work for Philology to do in its interpretation [Footnote: Such words, for instance, as [Hebrew script:],[Hebrew script:], [Hebrew script], used of different creative acts, may imply some difference of which we are ignorant. So again the uses of the words [Hebrew script], [Hebrew script:], and [Hebrew script:] for "man, " may have a bearing on some of those questions which now seem most perplexing.], but one thing seems certain—there is in it an absence of all detail. The facts to which it has reference are stated in the briefest and most simple manner, without the slightest reference to the means by which they were effected, or, apart from the question of the days, the time which was occupied in their accomplishment. When stripped of all that is traditional, and examined strictly by itself, the narrative seems greatly to resemble one of those outline maps which are supplied to children who are learning geography, on which only a few prominent features of the country are laid down, and the learner is left to fill in the details as his knowledge advances. Only in this case the details have already been filled in by the light of very imperfect knowledge, aided by a fertile imagination. These we must obliterate if we would restore the possibility of a faithful delineation, and we must be careful, in future, to avoid a similar error. We must put down nothing as certain which has not been conclusively shown to be so.

This last caution is specially needed at the present time, for, proud as we are of our advance in science, the amount of what is certainly known is probably very much less than we imagine. A great deal that was received as certain a few years ago, is now considered to be doubtful, or even recognized as a mistake and abandoned. This is especially the case with Astronomy, which seems to be almost in a state of revolution. Dependent, as it is almost entirely, upon mechanical and optical aid, every improvement and discovery in these departments changes its position, bringing to light new facts, and modifying the aspect of those which were previously known. The very basis of all astronomical calculations, the standard of time, is now no longer relied upon as invariable. It is suspected of a change resulting from a gradual retardation in the rate of the earth's rotation on its axis, produced by tidal friction. When the binary stars were discovered, the discovery was hailed as a proof of the universal prevalence of the law of gravitation. Later observations have thrown doubt upon that conclusion, as many pairs are known to exist, which, though they have what is termed a "common proper motion, " or are journeying through space together, have no relative motion, which they must show, if they were moving under the influence of their mutual attractions. The supposed simplicity of the solar system has given place to extreme complexity. A century ago, six planets, ten satellites, and a few comets, were supposed to constitute the whole retinue of the sun: now, instead of this, we have two groups of four planets each, the individual members of each group closely resembling each other in all points within our knowledge, while in all these points the groups differ greatly. Between these two groups lies a belt of very small planets, of which the 1st was discovered on the first day of the present century, and the 124th this year, and the number of known satellites has increased from 10 to 17. Add to this the meteoric groups, and their suspected connexion with certain comets, and the perplexing questions suggested by the Solar Corona and the Zodiacal light, and it will be seen that our knowledge is in a transitional state; that with so many problems unsolved, any apparent contradiction to the sacred record will require a careful scrutiny to ascertain that the grounds on which it is brought forward are well established.

Geology, so far as our present subject is concerned, stands upon a somewhat different footing. Though a much younger science than astronomy, it has one great advantage over it; the facts with which it has to do are for the most part discernible by the unaided senses, and it is therefore independent of instrumental help. Many changes have occurred in the views of Geologists, but in the main they have reference to processes [Footnote: Such, for instance, is the modification of the views of geologists as to the relative effects of "disruption" and "denudation" in determining the features of the earth's surface.] rather than to results, and it is the results with which we are chiefly concerned.

Physiologists have entered on the contest with the Bible on two different, and seemingly contradictory grounds. Some of them have maintained that the varieties of mankind are so distinct, that it is impossible they can all be descended from a single human pair, while others assert that not only all the varieties of mankind, but all the varieties of living beings are descended from a single progenitor. Between the advocates of these two systems there must be such an enormous difference as to the extent to which variation is possible, as to justify us in assuming that the fundamental principles of physiological science are not yet satisfactorily ascertained.

These are the three branches of science which come especially into collision with the Mosaic Record of the Creation. Of these Geology is the most important, because it is able to bring forward unquestionable facts which are in direct opposition to the traditionary interpretation Astronomy and physiology have little to object except theoretical views; the hypotheses of Laplace and Darwin. These, however, will have to be carefully considered. It will be necessary for us first to ascertain whether there really exists any such fundamental discrepancy between the record and ascertained facts, or theories so far as they are supported by facts, and stand on a probable footing, as should render all attempts at harmonizing them vain. If this is found not to be the case, we shall then be in a position to inquire whether modern discoveries afford us any really valuable light, and can assist us to form a somewhat more extended and accurate idea of the processes described by the sacred historian.

18

CHAPTER II.

DIFFICULTIES IN GEOLOGY.

The principal points on which there is a supposed discrepancy between the Mosaic Record and the discoveries of geologists are as follows: —

THE MOSAIC RECORD APPEARS TO ASSERT—

I. That the world in all its completeness, as it now exists, was moulded out of material in a chaotic state in six ordinary days. Geologists have ascertained, beyond the possibility of a doubt, that the process must have occupied countless ages.

II. That the first appearance of animal life was on the fifth of those six days. Geologists have discovered that animal life was in existence at the very earliest period to which they have as yet been able to extend their investigations.

III. That all living creatures are divided into two classes, and that the first of these classes was created on the fifth, the second on the sixth day; and that each class, in all its divisions, with the exception of man, came into existence simultaneously. Geologists trace the rise and increase of each class through a long course of ages.

IV. That death entered into the world through the sin of man. The very existence of fossils implies that it was the law of all animal life from the first.

V. That till the fall all creatures lived exclusively on vegetable food. Geologists have ascertained the existence of carnivorous creatures from a very remote period.

Besides these, there are some other supposed difficulties and inaccuracies of a less important character, which may be noticed, in passing, when the true meaning of the record is under discussion.

SECTION 1. THE DAYS.

The question of the days is beyond all doubt the most important of those which have to be discussed. On the one hand, the impression naturally left upon the reader of the first chapter of Genesis is that natural days are meant, and this impression is not removed by a cursory inspection of the original. On the other hand, if there is any one scientific belief which rests on peculiarly solid ground, it is the belief that the formation of the world occupied a period which is beyond the grasp of the most powerful imagination.

There is, indeed, some reason to think that the time claimed by geologists is somewhat exaggerated. Their views are in many cases based on the assumption that change is now going on, on the surface of the earth, as it did in all past time—that it is the same in character, in intensity, and in rate. But there are good reasons for supposing that almost all the causes which lead to change are gradually decreasing in intensity. The chief causes by which changes are brought about are the upheaval and subsidence of the earth's surface; the destructive agencies of wind, storms at sea, rain and frost; and the action of the tides. Of these, all but the last are directly dependent on the action of heat, and there is every reason to believe that the heat of the earth is in process of gradual dissipation. If this be the case, all those agencies which are dependent on it must

[Footnote: It is thought probable that this process is complete, or nearly so, in the moon. If this be the case, it is in all probability in progress in the case of the earth, though, owing to the much greater bulk of the latter, it occupies a longer period. —Lockyer, Lessons in Astronomy, p. 93.] be declining in intensity; but the rate of that decrease is unknown; it may be in arithmetical, or it may be in geometrical progression. It is, then, by no means impossible that changes, which now only become discernible with the lapse of centuries, might, at some past period of our globe's history, have been the work of years only. Nor is it at all probable that the present rate of change, which is assumed as the basis of the calculation, is known with any approach to accuracy. Exact observations are of very recent date; both the inclination and the means for making them are the growth of the last two centuries, and the changes which have to be ascertained are of a class peculiarly liable to modification from a variety of local and temporary causes, so that a very much longer period must elapse before we can arrive at average values which may be relied on as even approximately accurate.

Another circumstance, which seems to merit more attention than it has received, is the very frequent recurrence in Greek mythology of allusions to creatures which have been usually regarded as the creations of a poetic fancy, but which bear a strong resemblance to the Saurian and other monsters of the Oolite and Cretaceous formations. Of course, it is not impossible that these things may have been purely poetic imaginings; but, if so, it is very remarkable that such realizations of those imaginings should be afterwards discovered. It would seem much more probable that these legends were exaggerated traditions of creatures which actually existed when the first colonists reached their new homes, in numbers comparatively small, but still sufficient to occasion much danger and alarm to the early settlers, and to cause their destroyers to be regarded as among the greatest heroes of the time and the greatest benefactors of mankind. The Hindoo tradition of the tortoise on whose back stands the elephant which upholds the world, and the account of Leviathan in the Book of Job, seem to point in the same direction. [Footnote: For additional instances see Tylor's Early History of Mankind, p. 303.]

But, after all, the question is not one of a few thousands of years more or less, but of six common days, or many thousands of years. It may help us to arrive at a right conclusion on the subject if we endeavour to ascertain, in the first instance, whether there are any strongly-marked indications that the writer of the first chapter of Genesis did possess some accurate information on some points in the history of Creation which he was not likely to obtain by his own researches. For this purpose we will place in parallel columns the leading facts recorded by Moses, and a table of the successive formations of the rocks, abridged from the last edition (1871) of Sir C. Lyell's Student's Geology. This process will bring to light certain coincidences which may serve as landmarks for our investigation.

The Days.	THE ROCKS.
1. Creation of light.	
2. Creation of the Atmosphere.	

| 3. ⎧ The earth covered with water [implied].
⎨
⎩ Upheaval of land.
Creation of terrestrial Flora. | ⎧ Laurentian.
⎪ Cambrian.
⎨ Silurian.
⎪ Devonian.
⎩ Carboniferous. |

4. The sun and moon made "Luminaries." $\left\{ \begin{array}{l} \text{Permian.} \\ \text{Triassic.} \end{array} \right.$

5. Creation of birds and reptiles $\left\{ \begin{array}{l} \text{Triassic.} \\ \text{Jurassic.} \\ \text{Cretaceous,} \\ \text{Eocene.} \end{array} \right.$

6. $\left\{ \begin{array}{l} \text{Creation of land animals.} \\ \text{Creation of man.} \end{array} \right.$ $\left\{ \begin{array}{l} \text{Eocene.} \\ \text{Miocene.} \\ \text{Pleiocene.} \\ \text{Post Tertiary.} \end{array} \right.$

CONCURRENT EVENTS.

Laurentian: Upper Laurentian unconformably placed on Lower Laurentian, which contains Eozoon Canadense.

Cambrian: Traces of volcanic action. Ripple marks indicating land.

Silurian: Earliest fish.

Devonian: Earliest land plants.

Carboniferous: The coal measures. Peculiarly abundant vegetation. Earliest known reptile.

Permian: Foot-prints of birds and reptiles—with a few remains of the latter.

Jurassic: The first bird, and the first mammal. The age of reptiles.

Cretaceous: Reptiles passing away, mammalia abundant and of large size.

Post Tertiary: Human remains found only in the most recent deposits. In this table we see certain points of strongly-marked coincidence: —

1. The oldest rocks with which we are acquainted—the Lower Laurentian [Footnote: The age of granite is uncertain. —Lyell'a Student's Geology, p. 548.]—were formed under water, but had begun to be elevated before the next series, the Upper Laurentian,

were deposited. Ripple marks are found in the Cambrian group [Footnote: Ibid. p. 470], indicating that the parts where they occur formed a sea-beach, and, consequently, that dry land was in existence at that time.

2. The earliest fossil land plants as yet discovered are found in the Devonian series, and they gradually increase till, in the Carboniferous strata, they attain the extreme abundance which gave rise to the coal measures.

3. The age of reptiles. The earliest known reptile is found in the Carboniferous strata. In the Permian and Triassic groups the numbers gradually increase, till in the Lias, Oolite, and Cretaceous systems, this class attains a very great development both numerically and in the magnitude of individual specimens. During the same period the first traces of birds are found. The first actual fossil bird was found in the upper Oolite.

4. The age of mammalia. The first remains—two teeth of a small marsupial—were discovered in the Rhaetic beds of the Upper Trias, and a somewhat similar discovery has been made in beds of corresponding periods in Devonshire and North America. During the subsequent periods the numbers slowly increase, till in the Tertiary strata the mammalian becomes the predominant type.

5. The earliest traces of man—flint implements—are found in the Post Tertiary strata.

We have then in the Mosaic narrative five points which correspond in order and character to five points in the Geological record; and with reference to two, at least, of these points, we cannot imagine any cause for the coincidence in the shape of a fortunate conjecture, because, so far as we can tell, there was nothing apparent on the face of the earth to suggest to the mind of the writer the long past existence of such a state of things as has been revealed to us by the discovery of the Carboniferous and Reptilian remains. It seems then that Moses must have been in possession of information which could not be obtained from any ordinary source. But if he was thus acquainted with the order in which the development took place, there is nothing improbable in the supposition that he was not altogether ignorant of the length of time which that development required.

Let us suppose then that his knowledge did extend a little farther; let us suppose him to have been aware that each of the Creations which he describes was a process occupying many thousands of years—how could he have imparted this knowledge to his readers? What modification could he have introduced into his narrative, which without changing its general character, or detracting from its extreme simplicity, should have embodied this fact?

This amounts to the question: What words significant of definite periods of time were in use, and consequently at the writer's command, at this time? No language is very rich in such words; but in the early Hebrew they seem to have been very scanty. The day, week, month, year, and generation (this last usually implying the time from the birth of a man to that of his son, but possibly in Gen. xv. 16, a century) are all that we find. These in their literal sense were evidently inadequate. Nor could the deficiency be supplied by numerals, even if the general style of the narrative would have admitted their use, for we find in Genesis no numeral beyond the thousand. There was no word at all in early Hebrew equivalent to our words "period" and "season. " When such an idea was to be expressed, it was done by the use of the word "day, " either in the singular, or more commonly in the plural. Thus, "the time of harvest; " "the season of the first ripe fruit, " are literally "the days of harvest, " "the days of the first ripe fruit. " In Isaiah xxxiv. 8, the singular is used, and followed by the word year in the same indefinite sense. "It is the day of the Lord's vengeance, and the year of recompenses for the controversy of Zion. "

The only method then which was open to the writer was to make use of one of the words above mentioned in an extended sense, just as he used the word [Hebrew script] (earth) in several senses. But if one of them was to be employed, the one which he has chosen seems the best; not only because its use in that way was common, but because the brevity of the time covered by its natural significance would in itself be a hint of the way in which it was used. That which was impossible in a day might be possible in a year or a generation. The extended significance of the word would become apparent just in proportion as the time covered by its natural significance was inadequate for the processes ascribed to it.

An additional reason may, perhaps, be found for the choice of the word "day, " in the accordance of its phenomena with some, at least, of the processes which Moses describes—the dawn, the light slowly increasing to the perfect day, and then fading away gradually into night—these do seem aptly to represent the first scanty appearance, the gradual increase, and the vast development of plants, of the reptiles and of the mammalia, and in the case of the first two classes, their gradual passing away.

But if the word was thus employed in a figurative, and not in its natural sense, we may expect to find some indications in the context that this was the case. Such indications we do find. The fact that the work of Creation was distributed into days, is, in itself, significant. There is no reason to believe that in the opinion of the writer each day's work tasked to the utmost the power of the Creator. Moses was evidently as well aware as we are, that to Him it would have been equally easy, had He so willed, to call everything into instant and perfect being at a single word. Nor was the detailed description necessary to establish the foundation of all religion—the right of the Creator to the entire obedience of His creature For this the short recapitulation which (ch. ii. 4) prefaces the more detailed account of man's peculiar relation to his Maker would have been sufficient. Some purpose, however, there must have been for this more particular account which precedes the summary. We may trace two probable reasons. It brings before us the method of the Divine Working in the light of an orderly progress. But beside this, it is of infinite service to us, in enabling us more thoroughly to realize the Fatherly character and ever watchful care of our Creator. As far as that care itself was concerned, it was unimportant whether the work was instantaneous or progressive; but it was very important to us, in so far as it affected our conceptions of God, and of our relations to Him. For all our conceptions of God must rest ultimately on our self-consciousness; we can form no idea of Him except in so far as that idea is analogous to something which comes within the range of our own experience. Now to us and to our feelings there is a very wide difference between an act performed in a moment, and a work over which we have lovingly dwelt, and to which we have devoted our time, our labour, and our thought, for months or years. The one may pass from our mind and be forgotten as quickly as it was performed, but in the other we commonly feel an abiding interest. When therefore the great Creator is represented to us as thus dwelling upon His work, carrying it on step by step, through the long ages, to its completion, we find it far less difficult to realize that other truth, so precious to us, that His care and His tender mercies are over all His works, that the loving watchfulness which still upholds all, and provides for all, is but the continuance of that care which was displayed in the creation of all. Creation, Providence and Grace are blended together in one continuous manifestation of the Divine Wisdom, Power, and Love.

But for this purpose it is of little importance to us whether Creation is described as taking place in a moment, or in six ordinary days. If the division into six days indicates orderly progress and watchful care, we naturally expect to find the same indications in each of the subordinate parts. To our imperfect conceptions each single day's work would bear that same character of vast instantaneous action which seemed so undesirable. It would not help us to realize what it is so important that we should thoroughly feel. The very fact then that the history of Creation is divided into days carries with it a strong presumption that those days are not ordinary days.

In the 14th and following verses, when Moses is describing the formation of the heavenly luminaries, he is particular in mentioning that one part of their office was to "rule over the day and over the night, and to divide the light from the darkness. " Hence it is sometimes inferred that he was under a mistake in speaking of day and night at an earlier period. But such a mistake seems incredible. To suppose that Moses did not perceive that what he wrote in the 14th and following verses was incompatible with what he had written in the 4th and 5th verses, if such an incompatibility really existed, is to impute to him an amount of ignorance or carelessness which is at variance with the whole character of his writings from beginning to end. Instead of this it will be shown hereafter that, in all probability, his statements rested on a wide knowledge of facts. If then, under such circumstances, he uses the word "day" long before he comes to the formation of the sun, the natural inference is that he did so designedly—that it was his intention that his readers should understand that he was speaking of something very different from that natural day which is regulated by sunrise and sunset.

The way too in which he introduces the mention of the first and following days is apparently significant, though its full meaning is probably more than we can at present understand. In ver. 5 he carefully defines light and darkness as the equivalents of day and night; but in the next verse he passes over these words, and introduces two new ones, which he has not defined; these two words being as much out of place before the creation of the atmosphere as light and darkness are supposed to have been before the Creation of the Sun. And not only does he introduce two new words, but he introduces them in a very remarkable and, with our present knowledge, unaccountable manner. Had he said "And there was morning and there was evening, one day, " we should have found no difficulty in harmonizing; his words with what he had previously said concerning the evolution of light. But he first of all reverses the order, and then does not supply the natural termination to his sentence—"And there was evening and there was morning, "—"one night" would seem to be the natural conclusion; but instead of that we read, "there was evening and there was morning, one day. " Whatever farther significance then may be hereafter discovered in this remarkable statement, one thing at all events seems clear, that it was designed to call attention to the fact that the day spoken of was not a natural day. Probably certain stages in the progress of the work were indicated, which farther investigations may disclose to us. A few years ago such stages seemed to be discernible, but the continued progress of discovery has partly obliterated the supposed lines of demarcation. Still further discoveries may bring to light other divisions.

In the opening of the second chapter we are told that God rested on the seventh day from all His work, and His rest is spoken of in such a way as to carry our thoughts at once to the Fourth Commandment. In that commandment the duty of hallowing a seventh portion of our time is based on the fact that "in six days the Lord made heaven and earth, the sea and all that in them is, and rested the seventh day. " But the analogy entirely fails unless the days of the Creator's work bore the same proportion to the day of His rest which man's six days of labour bear to his Sabbath. Now we are expressly told in other parts of Scripture that the Divine Sabbath is not yet ended (Heb. iii. iv.), and we are led to infer that it will not end till He that sitteth upon the throne shall say, "Behold I make all things new. " If then the Sabbath of the Creator is measured by thousands of years—the whole duration of man upon the earth—it follows that the days of His work must have been of corresponding length.

One more indication, so strong that in itself it seems sufficient to decide the question, is to be found in the 4th verse of the second chapter. [Footnote: It is not unusual with critics of the German school to assert that this is an independent account of the Creation. But the assertion does not appear to have any valid foundation. The supposed grounds for it are well discussed in the "Speaker's Commentary, " vol. i. p. 23, and in "Aids to Faith, " Essay v., Sections 2, 4, 5. It has already been pointed out that the supposed variations in order rest entirely on the translation.] In that verse all that is ascribed to the six days in the preceding chapter is summed up as the work of a single day. If then the word is used in a natural sense in the first chapter, it is clearly used in an extended sense in the second chapter. But if it had been used in a natural sense in the first chapter, there would have been no need whatever for its use here. Its place would have been taken—and most appropriately—by the word [Hebrew script], a week, with which Moses was familiar (ch. xxix. 28; Deut. xvi. 10). Its use here would have connected the weekly division of time with the Creation, and as its presence would have been thus strongly significant, its absence is a no less significant indication that the six days spoken of in the preceding chapter are something very different from six natural days.

Three points, therefore, seem to be clear: —

1. However the chapter may be interpreted, there are in it coincidences with ascertained facts so marked that they cannot possibly be fortuitous. They prove therefore that Moses was in possession of some accurate information on the subject on which he was writing.

As we proceed with our subject we shall come upon many more indications of this, some of them exceedingly remarkable. It is therefore by no means improbable that he was acquainted with the fact, that the work which he was describing was one which had occupied a long series of ages.

2. Supposing that Moses was acquainted with all which has now been discovered by geologists, and that he was desirous of imparting that knowledge to his readers, the language which he has employed is the most appropriate that, under the circumstances, he could have chosen for the purpose. 3. The phenomena exhibited by the context indicate not only that he had this intention, but that he also intended that such of his readers as were competent to entertain the idea, should have sufficient indications to guide them to his meaning.

Whatever then may be the real significance of the "days"—a point which the knowledge at present in our possession seems insufficient to explain—it seems very clear that something very different from natural days is intended. And this is a sufficient answer to the objection which is founded on that interpretation. That there would be very many points which as yet we are unable fully to understand, has been already shown to be not only possible but probable; and among them it appears this question of the true meaning of the days must be left for the present. When we come to consider subsequently the great number of points in which harmony between the narrative and discovered facts is brought out on investigation, [Footnote: Chap. v.] we may well be content to leave many points unexplained till our knowledge is greatly increased.

SECTION 2. FIRST TRACES OF LIFE.

The second objection has reference to the relative antiquity of the various forms of life, of which we find traces in the successive strata of the rocks. If it be assumed that the apparent coincidences which have been pointed out between the Mosaic narrative and the geological records are real, and that the traditional interpretation is the true one, then we ought to find—

1. No traces at all of animal life below the Trias.

2. No traces of mammalia below the Cretaceous formation.

But the examination of the rocks leads to a very different result. Traces of life have been found, probably in the Laurentian, certainly in the Cambrian rocks. The earliest known fish is the Pteraspis, which has been discovered in the upper Silurian formation at Leintwardine, in Shropshire. The first member of the reptilian order, Archegesaurus, occurs in the coal measures; and the first traces of a mammalian—two teeth—occur at the junction of the Lias and Trias. In every case, then, we meet with traces of life at a period long anterior to that at which we should naturally expect them.

In order to ascertain the real weight of this objection we hare to investigate two points: —

1. What are the animals to which the Mosaic Record refers?

2. What does it really tell us about the creation of those animals?

1. It is commonly assumed that all living creatures are comprehended under the terms used in describing the work of the fifth and sixth days. But a more careful examination shows that there is no real ground for this assumption. The first point which presents itself is the omission of the Hebrew word for fish, [Hebrew script], in

the account of the fifth day—an omission the more marked, because the word does occur in vv. 26, 28, in which dominion over all living creatures is granted to man. The two words which are used in ver. 21 are [Hebrew script] from [Hebrew script], to stretch out, to extend, and [Hebrew script], from [Hebrew script], identical with [Hebrew script], to trample with the feet. The description then points us to animals of great size, especially length, which trample with the feet. "Great sea- monsters, " Gesenius calls them. These words clearly indicate the Saurian and allied tribes of reptiles; and when we turn to the rocks we find the remains of these creatures occurring in great numbers, precisely at the point which Moses assigns to them.

Again, in the account of the sixth day, three classes of animals are mentioned; but we have no means whatever of ascertaining what kinds of animals were comprehended in these three classes, or whether they included all the mammalia then known to the Jews; much less then are we justified in inferring that they comprehend all mammalia that were then, or ever had been in existence.

But it may perhaps appear strange, that the account of the Creation of living beings should be of such limited extent, embracing only reptiles, birds, and mammals. A little consideration, however, will remove this apparent strangeness. We should, perhaps, naturally expect to have some notice of the first appearance of animal life; but from the circumstances under which Moses wrote such a notice was simply impossible. The lowest and simplest form of life with which we are now acquainted is the Amoeba Princeps, a minute particle of jelly-like substance, called sarcode—scarcely larger than a small grain of sand—and with no distinction of organs or limbs. [Footnote: Carpenter, The Microscope and its Revelations, p. 428.] The oldest known fossil, Eozoon Canadense, is of a class but little above this— the foraminifera; we may therefore deem it probable that life began with some form not very unlike the Amoeba. How could the formation of such a creature have been described to the contemporaries of Moses? They could have had no idea of its existence. To describe the first beginnings of life then, was, under the circumstances, an absolute impossibility. But if a part only of the long series of animal life could possibly be noticed, the determination of the point at which he should first speak of it would be left to the writer, guided as he would be by considerations of the object for which, and the persons for whom, he wrote, which we must necessarily in our position be unable duly to estimate. All that we are entitled to expect is that the account, so far as it extends, should be in accordance with facts.

The next point to be ascertained is, "Does the Mosaic Record intimate that the creations of reptiles on the fifth, and of mammals on the sixth days were entirely new creations, i.e. that no creatures of these classes had existed before? " There is no direct assertion to this effect; it is only an inference, though a natural one, when we consider the circumstances under which it was drawn. When, however, we turn to the original we find the 20th verse worded in a way which seems designed to avoid the suggestion of such an inference. Literally translated it is, "Let the waters swarm swarms, the soul of life. " Such creatures then may have existed before, but not in swarms. And in the account of the sixth day, as has been already noticed, three forms of mammalia are specified, and we have no knowledge as to the varieties included in these three forms. Nor is there here any intimation that it was the first creation of such animals. The greater part of the earlier fossils belong to the Marsupialia and Mouotremata, and we have no reason to believe that these classes have existed in historic times in Europe, Asia, or Africa. They are now confined (with the exception of the opossums, which are American) to Australia. They were therefore entirely unknown to the Jews, and in consequence necessarily omitted in a document intended for their use.

What has been said with reference to reptiles is also applicable to birds. The first traces of them are found in the ornithichnites of the new red sandstone, and the first fossil—Archaeopteryx, in the Solenhofen strata, belonging to the Oolite. From the nature of the case the remains are necessarily scanty, since birds would be less exposed than other animals to those casualties which would lead to their preservation as fossils, but enough traces have been found to show that in the period corresponding to the fifth day they were very numerous, and attained in many instances to a gigantic stature. A height of from ten to twelve feet was not uncommon.

When, therefore, we notice that the fifth and sixth days correspond to two periods, in the first of which reptiles and birds, and in the second mammalia, were the prominent types, the words of the sacred historian seem to have an adequate interpretation in that fact. There is no contradiction between the two records. Moses describes but a very few of the facts which geology has brought to light, but those few facts are in exact accordance with the results of independent observation. The acts of Creation of which Moses speaks correspond to remarkable developments of the orders of animals to which he refers. To have noticed the time of the appearance of the first individual member of each class, as distinguished from the time when that class occupied the foremost place in the ranks of creation, would have been inconsistent with the simplicity and brevity of the narrative, while it would have been unintelligible to those for whom the narrative was intended, since these primeval types had passed out of existence ages before the creation of man. It is, however, noteworthy, that the first appearances of the several orders follow precisely the same arrangement as the times of their greatest development.

SECTION 3. SIMULTANEOUS CREATION.

This objection may be very briefly disposed of, though it appears to be one which has made a very deep impression on Mr. Darwin. [Footnote: Origin of Species, p 1, &c.] It is entirely an inference drawn from the old interpretation of the six days. While that interpretation was received it followed, as a necessary consequence, that the creation of all kinds of plants on the third day, and of reptiles, birds, and mammalia on the fifth and sixth days respectively, must have been simultaneous. But if that interpretation is proved to be untenable, the inference drawn from it falls to the ground. The language of the narrative seems to point in an opposite direction. There is one instance in the chapter in which the words used seem to point to an instantaneous result. "And God said 'Let light be' and Light was, " though in this case the words probably have a further significance, which has been brought out by the discovery of the nature of light. But in these three cases the command is first recorded, with (in two cases) the addition "and it was so, " and then the narrative goes on to speak of the fulfilment of the command, as if the command and its fulfilment were distinct things.

SECTION 4. DEATH. CARNIVOROUS ANIMALS.

These two objections may advantageously be considered together, since the fifth is in a great measure, though not entirely, dependent upon the fourth. For if death, in the common sense of the word, was unknown till the fall of Adam, it follows as a necessary consequence that no carnivorous creatures could have existed before that time. On the other hand, it may be considered as the natural death of large classes of animals to be devoured by the carnivora; so that if there were no carnivorous animals prior to the Fall, one of the avenues to death, at all events, had not been opened.

There is really no ground at all for the first of these objections in the actual history of Creation. It is only when the threat held out to Adam (ii. 17) is viewed in the light of St. Paul's comment upon it (Rom. v. 12; viii. 20) that the supposition can be entertained. This, then, is the real foundation of the difficulty.

But, first of all, there is no reason to suppose that St. Paul's words refer to any death but that of man. Now, it may well have been, that although man, having a body exactly analogous to those of the animals, would naturally have been subject, like them, to the ordinary laws of decay and death, yet in the case of a creature who possessed so much which raised him above the level of the lower animals, there may have been some provision made which should exempt him from this necessity. That this was the case appears probable from the mention made in the narrative of the Tree of Life. We have no intimation whether the action of the fruit of this tree was physical or sacramental, but that, in one way or other, it had the power to preserve man from physical death seems almost certain from the way in which it is spoken of after the Fall (iii. 22-24). But the mention of the Tree of Life leads to the inference that the case of Adam was entirely exceptional.

In the next place, it does not seem probable that that dissolution of the body which was the natural lot of all other animals was the whole, or even the chief part, of the evil consequence of Adam's fall. That it was included in the penalty seems probable, but it only constituted a comparatively unimportant part of that penalty. The threat was, "In THE DAY that thou eatest thereof thou shalt surely die, " and we cannot doubt that the Divine words were exactly fulfilled, though Adam's natural death did not take place for many hundred years. But the guilty creatures, covering their nakedness with fig-leaves, crouching among the trees of the garden in the vain hope of hiding themselves from the face of their Maker, who were to transmit an inheritance of sin and shame and misery to their yet unborn posterity, were surely very different beings from those whom the Creator but a short time before had pronounced "very good. " The true life of the soul was gone; the image of God defaced. This was the real, the terrible death. If death in its full sense means nothing more than the dissolution of the body, our Lord's words,

33

"He that liveth and believeth in Me shall never die, " have failed of their fulfilment. That promise has been in force for more than eighteen centuries, and yet no case has occurred of a Christian, however holy he may have been, or however strong his faith, who has escaped the universal doom. The Church of the Patriarchs could point to an Enoch, the Jewish Church to an Elijah, who were exempted from the universal penalty; but Christianity can point to no such exemption, nor does she need it. To her members, to die is to sleep in Jesus; to be absent from the body is to be present with the Lord, for the penalty of death is cancelled.

Though, then, it seems by no means improbable that Adam, if he had not fallen, would have been exempt from the dissolution of the body, yet this is not absolutely certain, and even if it were certain, his case would be an exceptional one: no inference as to the immortality of the animal creation could have been drawn from it.

The supposition that all animals prior to the fall lived entirely on vegetable food rests partly on this groundless inference, and partly on the Divine Words recorded in verse 30: "And to every beast of the field, and to every fowl of the air, have I given every green herb for meat. " But it is important to notice that these words are not recorded as addressed to the animals, like the command to be fruitful and multiply. Had this been the case, any omission to mention the flesh of other animals, might have been looked upon as significant. Instead of this they are addressed to Adam, and they follow other words in which the same things are assigned to Adam for his food. They come then in the form of a limitation to the rights granted to Adam, rather than of a definition of the rights of the lower animals. Adam was to have the free use of every green herb, but he was not to account himself the exclusive owner of it. The beast of the field and the fowl of the air were to be co-proprietors with him; they were to have the use of it as freely as himself; but that they were to be restricted to the use of vegetable food nowhere appears. Accordingly we know that carnivorous creatures have existed from the first, and that though to a superficial observer this may appear a cruel arrangement, yet in reality it is a most merciful provision, by which aged, weak, or maimed animals are preserved from the agonies of death by starvation.

We may conclude then that there is no real contradiction between the conclusions at which Geologists have arrived, and the words actually made use of by Moses, but that all such supposed contradictions have arisen from meanings being attached to those words, which, though possible or even probable, were not the only possible meanings. When the difficulty has been suggested, and the words have in consequence been more closely examined, it appears that they are capable of an interpretation in strict harmony with every fact which Geologists have as yet discovered, and that in many cases there are not wanting indications that the writer intended them to be thus understood.

CHAPTER III.

DIFFICULTIES IN ASTRONOMY.

These objections, so far as they are based or supposed to be based on ascertained facts, are very few and insignificant. The chief of them are as follows: —

1. Moses describes light, and the division of night and day as existing before the Creation of the Sun.

2. Moses describes the firmament as a solid vault.

3. Moses speaks of the stars as created on the fourth day, only two days before Adam, whereas astronomers have asserted that many of them are so distant that the light by which we see them must have been on its way ages before Adam was created.

That part of the first objection which refers to the existence of light prior to the creation of the Sun, appears so extremely childish that it might have been thought unnecessary to notice it, had it not been solemnly propounded in such a work as "Essays and Reviews. " [Footnote: Page 219] Anyone who is in possession of a telescope of but moderate power may satisfy himself of its futility on any starlight night. He has only to turn his telescope to one or two of the more conspicuous nebulae; the Great Nebula in Orion, for instance, or the Ring Nebula in Lyra, and his eye will receive light which has not come from any Sun, for it is a well- ascertained fact that these nebulae are nothing but vast masses of incandescent gas. And this objection is singularly inappropriate in the mouth of the opponents of the Mosaic Record, inasmuch as the Nebular hypothesis is with them the favourite method of accounting for the present state of things. The view which they bring forward as an alternative to the Mosaic account assumes the very state of things which, when, alleged by Moses, they denounce as impossible. The other part of this objection, which refers to the division of day and night, will be more advantageously discussed when we come to consider the actual accounts of the first and fourth days' work. It will then appear probable that the statements which Moses has made on this subject, instead of being indications of ignorance, are the result of a profound knowledge of the subject on which he was writing.

Next, it is alleged that Moses describes the firmament as a solid vault. [Footnote: Essays and Reviews, p. 220.] "The work of the second day of creation is to erect the vault of heaven, which is represented as supporting an ocean of water above it. " That the Greek and Latin translations in this place do seem to imply the idea of solidity seems indisputable; and from the Latin the word "firmament" has passed into our own language. But there is no reason to think that the Hebrew word has any such meaning. It is derived from a root signifying "to beat out—to extend. " [Footnote: May not this root, [Hebrew script], have some connexion with [Hebrew script], "to be light, " from which is derived the Aramaic "Raca" of Matt. v. 22?] The verb is often applied to the beating out of metals, but not always. It is a new doctrine in etymology, that the meaning of a verbal noun is to be deduced from the nouns which often supply objects to its root, instead of from the meaning of the root itself. But even if it can be shown that the word did originally involve such a meaning, that would be nothing to the purpose. It would only be in the same case with a vast number of other words, which, though etymologically untrue, are habitually used without inconvenience, because they do convey to the minds of others the idea which we intend to convey, their etymology being lost sight of. Probably, the very persons who bring forward the objection do sometimes use the word "firmament, " though they know the error which is involved in it. Nor would they be any more accurate if they substituted for it the Saxon word "heaven, " since that also involves a scientific inaccuracy. The word used by Moses was the commonly recognized name for the object of which he was writing; and no objection to his use of it can be maintained, unless it can be shown that in using it he rejected some other word equally intelligible to all, and which was at the same time etymologically correct. But there is no ground for the assumption that any such word existed in the time of Moses or at any subsequent period.

The third objection, of course, ceases to have any force if the days of creation are no longer regarded as natural days. But the objection is in itself, apart from this condition, of no consequence whatever. For, in the first place, it is by no means certain, or even probable, that the stars referred to in the fourth day's work are the fixed stars. The Hebrew has no word for planets as distinguished from the fixed stars, although, as we know for certain, the difference between the planets and the fixed stars was recognized from a very early period. In every case, then, the context must determine the sense to be given to the word. In this case, the fact that these stars are mentioned in connexion with the sun and moon, combined with our knowledge that the planets, like the moon, are dependent upon the sun for their light, would lead us to infer that they are meant.

But even if the fixed stars were meant, the objection would be no longer tenable. It rests on certain estimates as to the supposed distances of the fixed stars and star clusters, which were formed by the late Sir W. Herschel from what he designated the "space-penetrating power" of his telescopes. Starting with the assumption that the stars were of tolerably uniform size and brilliancy, and that the difference in apparent brightness was the result, and therefore a measure of their distances, he proceeded to apply the same process to the star clusters, which, even in a fair telescope, present only the appearance of faint nebulous spots of light, but are resolved into clusters of stars by more powerful instruments. In many cases, he found that a certain proportion existed between the telescopic power by which a cluster was first rendered visible, and that required for its resolution, and by this means he formed what he considered a probable estimate of its distance. Other clusters there were which only became visible in his most powerful telescopes, and which, therefore, he could never succeed in resolving. These he placed at a still greater distance, and from this estimate he deduced the conclusion that their light must have been in some cases as much as 60,000 years in reaching the earth.

But the whole foundation on which this long chain of inference rested has now been shown to be evanescent. In the first place many of his irresolvable nebulae have been proved by the spectroscope to be true nebulae—masses of luminous gas, and not star clusters at all; and, in the next place, the actual distances of a few of the fixed stars have been approximately ascertained, and it is proved beyond all doubt that the different degree of brightness exhibited by different stars is no test at all of their distance. Of all the stars in our hemisphere whose distance has thus been measured, the nearest to us is one which can only just be discerned by a practised eye on a favourable night, 61 Cygni, whilst the most brilliant star visible in England, Sirius, is at a considerably greater distance. The most competent judges estimate the magnitude of Sirius as about one thousand times that of the sun [Footnote: Mr. Proctor in Good Words, February, 1872.]. In addition to this, many stars of very different magnitudes are found to be related to each other in such a way as to show that they are in actual, and not merely in optical proximity. The clusters which were formerly supposed to consist of large stars at enormous distances from us, are now, upon very solid grounds, believed to be formed of much smaller stars, at much more moderate distances, so that it is very improbable that there is any object visible in the heavens whose light has taken so much as 6000 years, instead of 60,000 years to reach us.

THE NEBULAR THEORY.

We come now to the consideration of the Nebular Theory of Laplace, in so far as it is opposed to the Mosaic account. It must be remembered that, after all, this is only a theory. Even if it could be satisfactorily established, it would only point out a way in which this world MIGHT have been formed. That it could not have been formed in any other way is an independent proposition, in support of which no single argument has ever yet been brought forward. There may be a greater or less probability that the earth was formed in this particular way, that probability depending on the extent to which the theory accounts for observed facts. This it does in many cases, and it has in consequence been accepted AS A WHOLE by many scientific men, as a substitute for the Scriptural account. As will be seen hereafter, there are strong reasons for admitting it as a supplement to the brief account given by Moses; but our business now is to ascertain, whether it has any just claim to be received instead of that account.

The theory seems to have been suggested by certain speculations of Sir W. Herschel. In his telescopic examination of the Nebulae and star clusters, he found that in a great number of cases, when a nebula was rendered visible by a certain amount of telescopic power, it would be resolved into separate stars by a telescope of a little higher power. But there were some nebulae, visible in very small telescopes, or even discernible with the naked eye, such as those in Orion and Andromeda, which could not be resolved even by his great four-foot reflector, the largest telescope that had then been constructed. And these nebulae exhibited a great variety of forms. Some of them were vast shapeless masses of faint light; others, which he designated "planetary" nebulae, exhibited a regular form—a circular disc more or less clearly defined, often brightest in the centre. Others seemed to be intermediate between these two classes. Hence he was led to the idea that these were worlds in the process of formation, and that their varying forms indicated varying stages of that process.

This suggestion was eagerly adopted by the members of the French Academy, who were at that time on the look-out for anything which they thought would help them to account for the existence of the world, while they refused to acknowledge a Creator. It was taken up by one of their number—Laplace—a man who stood in the very foremost rank as a mathematician and physical astronomer, and moulded into shape by him. [Footnote: There is a very full account of Laplace's hypothesis, extracted from the works of Pontecoulant, in Professor Nichol's System of the World, pp. 69—86.]

He assumed, that the Solar System existed at the very earliest period as a shapeless nebula, a vast undefined mass of "fire- mist; " that at some time or other the separate particles of this fire-mist began to move towards their centre of gravity, under the influence of their mutual attractions, and thus assumed a spherical shape; that by some means or other a motion of rotation was originated in this spherical mass, which increased in rapidity as the process of condensation advanced. The effect of this rotation would be a flattening of the sphere; the equatorial diameter would increase while the polar diameter, or axis of rotation, diminished; and when the centrifugal force thus produced had reached a certain point, a ring would detach itself from the equator, but would continue to revolve about the common centre. He supposed that a succession of rings were thus thrown off, which finally broke up and accumulated into one or more spherical masses, forming the planets and their satellites, while the remainder of the original sphere was condensed into the sun. The planets and their satellites would continue to revolve about the centre as the ring from which they were formed had done, while the different original velocities of the particles of which they were formed, some having been in the outer, some in the inner part of the ring, would cause them also to rotate on their axis. As the condensation advanced, the heat which had originally existed in the "fire-mist" would be condensed also, so that all the masses when formed would be in an incandescent state, but the planets and their satellites being comparatively small would soon cool down, while the sun, owing to its greatly superior bulk, still retains its heat.

There is no doubt much to be said in favour of this theory, which may be more advantageously considered hereafter, when we shall have to consider it as supplementary to the Mosaic account. At present we are only concerned with it as it claims to stand alone, and to be accepted as a substitute for that account. Viewed in this light, as a substitute for a Creator, as showing us how the universe might have come into existence spontaneously, it utterly breaks down in three points.

1. It gives us no account whatever of the origin of matter, but assumes that it was already in existence at the time from which the theory takes its point of departure. But some account of it must be given. Either it was created by some higher power, or it was eternal; for the idea of its being self-originated is manifestly untenable. If it was created, there is an end of the theory—the act of creation assumes the existence of a Creator; and the only question left is, whether that Creator did more or less. But the very object of the theory was to dispense with the existence of a Creator. This alternative, then, it must reject, and there is nothing left but to fall back upon the other, and to assume that it existed from all eternity. But it is certainly not less difficult to us to conceive the possibility of inert matter being self-existent and eternal, than it is to recognize the existence of an eternal and all-powerful Spirit. Our own consciousness helps us to realize the possibility of the existence of an Eternal Mind, and of the exercise of power by that mind; but we have nothing to help us to a conception of self-existent matter.

In addition to this, the idea of eternity precludes from its very nature the idea of possible change. If there is change there must be the distinction of before and after, and so of the succession of existence, which involves the idea of time. That which is subject to change, and this theory assumes a change in the condition of matter, cannot be eternal.

2. The next failing point is, that this theory assumes a change, of the origin of which it can give no account. The assumption is, that matter which had existed from all eternity, or for an indefinite time, in a state of perfect rest, suddenly began to move towards its centre of gravity. A body, or a system of particles, can remain at rest only under one of two conditions. Either it must be acted on by no force at all, or all the forces by which it is acted on must be in perfect equilibrium. If matter existed under the first of these conditions, whence did the force suddenly emanate? Force cannot be self-originated any more than matter. But if the other alternative be adopted, how was the equilibrium disturbed? It is a fundamental axiom of mechanics that "a body (or system of bodies) at rest will continue at rest till it be acted upon by some external force. " But the theory supplies no such external force, for it could only originate in

that which the theory ignores—the will and power of some intelligent Being.

3. The third defect is, that the theory does not give any satisfactory account of the origin of the motions of rotation and revolution. Laplace does not attempt this. He simply assumes that a motion of rotation was set up somehow; but many of his followers, perceiving that the theory broke down here—though they passed the other two defects unnoticed—have attempted to supply the deficiency in this point. Some have attempted to account for this motion by analogy. It has been suggested that it was of the same nature, and produced by the same causes, as the vortex which is formed when a vessel full of fluid is emptied through an orifice in its bottom. Pontecoulant, in his account of the theory, enters more into detail. He assumes that in the process of agglomeration large bodies of matter impinged obliquely on the already formed mass, and so imparted to it a motion of rotation.

A consideration of the mechanical conditions of the problem will show the unsoundness of Pontecoulant's views. It is of course assumed that the forces by which this rotation is said to have been produced are identical in their character with those with which we are familiar, for the introduction of any force peculiar to that time would be equivalent to an admission of a directing power. The following propositions then seem unquestionable: —

1. The nebula must be considered as a system of particles acted on by their mutual attractions, and by no other force.

2. When two particles of matter, a and b, attract each other, it is a fundamental principle of mechanics, (commonly known as the "Third Law of Motion") that whatever amount of momentum is produced in a, an equal and opposite momentum must be produced in b. Hence if the mutual action remain undisturbed, the two particles will approach each other and finally meet. On their union, the two momenta being equal and opposite will neutralize each other, and there will be no tendency to produce motion of any kind.
3. The same law will hold good with reference to any number of particles, and therefore with reference to the supposed nebula. Every single particle will produce a certain momentum in each of the other particles, and at the same time will have impressed upon it by each of the other particles an equal and opposite momentum. Hence when all the particles are collected into a single mass, each individual

momentum will be balanced by an equal and opposite one, and there can be no resultant motion.

The analogy from fluids flowing through an orifice fails, because—

1. The particles of the fluid are acted on by forces other than their mutual attractions, and in many cases affecting them unequally, e. g., friction against the sides of the containing vessel and the orifice.

2. Because the orifice is not a point, but a finite area, and consequently the particles of the fluid are acted on by forces which do not pass through the same point.

Considered then as a substitute for the action of an intelligent Creator, Laplace's theory utterly breaks down in three points, which, as they will have to be referred to hereafter, it is well to recapitulate.

1. It does not account for the origin of matter.

2. It does not account for the emergence of the force of attraction.

3. It does not give a satisfactory account for the motion of rotation.

CHAPTER IV.

DIFFICULTIES IN PHYSIOLOGY.

The third science which is supposed to come into collision with the Mosaic Record is Physiology. Here, however, we meet with no objections which rest upon ascertained facts, as in the case of geology. We have only to do with theories. All that can be brought forward is merely matter of opinion or theory—such theory resting indeed on a foundation of ascertained facts—but being in itself a mere inference more or less probable from those facts. Even if it were proved to be a true account of the causation of those facts, it would be by no means certain that other facts, however similar, might not have had a totally different origin.

At one time it was very confidently asserted, by many eminent physiologists, that the differences between various branches of the human race were so great, that it was impossible that all should have descended from the same original stock. Probably this opinion is still maintained in some quarters, but of late years views of a diametrically opposite character have been brought forward, and very ably advocated. In proportion as these views are admitted to have in them an element of truth, the importance of the older objection is diminished. It will therefore be unnecessary to dwell upon it. This new view is, that not only all branches of the human race, but all living beings now existing, or that have ever existed on the face of the earth, are descended by the process of "evolution, " carried on under what are designated as "natural laws" from some one variety, or small number of varieties of living creatures of the lowest type.

This theory, like that of Laplace, had its origin among the French Academicians, at the close of the last century. Its author was La Marck. According to his view the simplest form of animal life, the "monad, " was spontaneously developed by some unknown process. From this monad higher forms of animal life were produced, and the course of development was continued till it finally culminated in man. But it does not appear that La Marck suggested any means by which the various stages of development were brought about, and the view attracted little attention. Some thirty years ago it was revived by an anonymous writer, in a work called "Vestiges of Creation. " In this work the idea of spontaneous generation was repudiated. The original monad was supposed to have derived its existence from an act of Creative Power, and to have been then left to work out its own development, by virtue of powers originally implanted in it. All its variations and advances were supposed to be the result of the will and efforts of the creature acting through many generations. Thus the desire and attempt to walk ended in the development of legs, while wings were the final result of its efforts to fly. It was felt, however, that this was by no means a satisfactory account of the state of things, and so the work, though it produced a great sensation at the time, has now been almost entirely forgotten.

Latterly, however, the theory has found a far more able advocate in the person of Mr. Darwin, with whose name it has been popularly identified. By his indefatigable labours a vast variety of facts have been collected and skilfully arranged, to show that all the varieties of life may be satisfactorily accounted for by the continued action, through a long course of ages, of certain natural causes, with the results of which we are familiar, and of which intentional use is continually made by man. Mr. Darwin does not deny the existence of a Creator, but the tendency of his arguments is to prove that His interference was limited to the single act of original Creation; and that from the moment of its creation the world has been a sort of automatic machine, producing its results without any interference from any higher power.

The theory taken as a whole comes into contact with the Mosaic Record in three points: —

1. As it assumes the possibility that life may be self-originated.

2. As it indicates a mode of procedure different from that given by Moses.

3. As it requires unlimited time.

Of these the last is already disposed of, when the narrative is shown to be capable of an interpretation in accordance with it. The first requires only a brief notice; but the second must be carefully investigated, to separate ascertained truth from inferences which have no sufficient foundation.

The theory of spontaneous generation rests almost entirely upon assumptions. Its only semblance of support from facts is derived from certain experiments of a very unsatisfactory character, which are said to have resulted in the production of some of the lowest forms of animal life. These experiments have been by no means uniformly successful. One or two experimenters have thought that they have succeeded, but not uniformly, while the same process, repeated by men whose scientific and manipulative powers are universally recognized, has never once resulted in any seeming development of life. Even if, however, they had been uniformly successful, there would have been great reason to doubt whether the apparent success was not really a failure—a failure in the precautions necessary to exclude all germs of life from the matter experimented upon. For the lower forms of life are excessively minute; and their germs—eggs, seeds, or spores—must be far smaller. It is known that these are constantly floating in the atmosphere, though, owing to their extreme minuteness, the fact can only be ascertained by the most skilful investigation. And the lower forms of animalcules have a singular tenacity of life; they can pass unharmed through processes which would be fatal to creatures of higher organization. One variety is known to survive entire desiccation; another lives upon strychnine; others bear without injury great extremes of heat and cold; and if this is the case with the mature creatures, it is probable that the germ possesses still stronger powers of vitality. If one acarus can live upon strychnine, then it is not impossible that mineral acids should be harmless to others; the germs might be carried through sulphuric acid in air without coming into contact with the acid, as air would pass through in bubbles, in the centre of which they might be suspended; or if like the diatomaceae, they were coated with silex, they might come into contact with it and resist its action. Thus one of the precautions commonly taken is not certain in its action, and the same might be shown to be true of the others. The theory of spontaneous generation is, in fact, generally repudiated by Evolutionists, and cannot therefore be taken as a starting-point.

We come then to the theory of Evolution with which Mr. Darwin's name is associated. This theory asserts that all the varieties of animal life now existing on the earth, however widely they may differ from each other, are in reality derived from one, or a very few original types; and that in this general statement the human race is to be included. This theory rests upon the following admitted facts.

1. There are not, as was at one time commonly supposed, broad and distinct lines of demarcation between the different varieties of

animals and plants. Our increasing knowledge of zoology has brought to light the fact that one species shades off into another by almost imperceptible gradations. As we go back in the fossil records of animal life in the past, we find that the species now existing, while they are closely allied to correspondent species of an earlier period, are scarcely ever identical with them, and that the few cases of identity which do occur, are limited to the most recent rocks. Either then the old species must have perished, and new ones, similar but not identical, must have been created to take their places, or there must have been a process of gradual change, by which the present species have been derived from their predecessors. In one or two cases fossils have been found which combine, to some extent, forms which are now found in distinct species, as if the process of variation had proceeded in distinct lines from a common source.

2. No two animals of any class are exactly alike in all points. Each has its individual peculiarities, and in some cases these peculiarities are strongly marked.

3. Man has been enabled, to a certain extent, to make use of these individual peculiarities, and by means of them to produce great varieties in the breeds of domesticated animals. This has been sometimes done unconsciously through a selection influenced by other motives, and then the process has been very slow; but latterly intentionally, with a view to the production of improved breeds, and whenever this has been the case, changes of considerable extent have been rapidly produced. By carefully selecting the animals to be paired, any desired modification can generally be produced in the course of a few generations. This is exemplified in the numerous and increasing varieties of the breeds of almost all domestic animals and birds.

The theory of Evolution then suggests that the same processes which are employed by the cattle-breeder have been in operation through untold ages. For the intention and care of the human agent, Mr. Darwin substitutes two principles; one designated as "Natural Selection, " the other as "Sexual Selection. " For their full development he claims unlimited time. The ground on which the Process of Natural Selection is maintained is as follows: —

It has been already noticed that no two individuals of the same kind are exactly alike in all respects; each individual has some peculiarities, generally very trifling, but sufficient to distinguish it from all other individuals. Some of these peculiarities will probably be such as to be of some service to the individual in the struggle of life; they will assist it in procuring food, or in resisting or escaping from its natural enemies, while on the other hand the peculiarities of other individuals will be prejudicial to them in these ways. The consequence will be that a larger proportion of those having favourable peculiarities will survive and propagate their kind; their offspring will inherit the peculiarities of their parents, and reproduce them in various degrees. The same process will then be repeated, and thus from generation to generation the peculiarity will be increased, till at last it is sufficient to mark out, first a new variety, then a new species, and so on. This process then, continued through a long course of ages, was at one time considered by Mr. Darwin sufficient to account for all the varieties of living creatures now existing, or that have existed in past ages. But he has more recently satisfied himself [Footnote: Descent of Man, vol. i p. 152.] that there are many phenomena which are not satisfactorily accounted for by this principle, since many of the specific differences of animals are found to exist in matters which, cannot directly promote their success in the struggle of life. Such, for instance, are the brilliant colours which are found, especially among the males, in many species of birds. These he proposes to explain by the supplementary theory of "Sexual Selection. " His suggestion is that these peculiarities are in some way attractive to animals of the opposite sex, so that the individuals in which they are most strongly developed are more successful than others in obtaining mates, and that in this way the peculiarity is gradually fixed and increased.

By these two processes, then, Mr. Darwin supposes that all the differences now existing among animals have been produced and perpetuated; and not only that, but that man also is the result of similar processes, acting through a very long period; that the progeny of certain "anthropomorphous apes" have, by slow degrees, risen in the scale of being above their progenitors; that all our faculties, intellectual and moral as well as physical, differ from those possessed by lower animals in DEGREE only, and not in KIND, [Footnote: Descent of Man, chaps, ii. -v.] so that man has arrived at his present state by what may be termed purely natural processes, without the intervention of any external power.

In considering these theories, our attention must first be directed to some defects which appear to weaken the whole course of the argument; and then we may consider the peculiar difficulties in the way of the processes of natural and sexual selection; and the grounds for the belief that man is in possession of something entirely different in KIND from any faculty or power possessed by any lower animals, which could not therefore be derived by inheritance and improvement.

The first thing which strikes us in Mr. Darwin's works is that, from time to time, he betrays a sort of latent consciousness that his theory is insufficient; that the processes to which he ascribes such vast results are not quite adequate to the purpose, but that they need in some way to be supplemented. Every now and then recourse is had to some law—some unknown cause—which must co-operate in the production of the results he is considering. In spite of the apparent care which he has taken to guard against it, he is continually betrayed into a confusion between the two senses in which the word "law" is employed. In its proper significance, law is an expression of the will of an intelligent superior, enforced by adequate power. In this sense the law may be considered as an efficient cause. The combination of will and power is an adequate cause for any result whatever. But Mr. Darwin expressly excludes this sense of the word, in a sentence which seems to involve a self-contradiction. "I mean by nature only the aggregate action and product of many natural laws, and by law only the ascertained sequence of events. " [Footnote: Plants and Animals under Domestication, vol. i. p. 6.] Law, in this sense, then, is simply the statement of observed facts, and as such can have no action at all. It asserts that certain phenomena do uniformly follow each other in an ascertained order; but it gives us no information whatever as to the cause of those events, or the reason why they do thus succeed each other. But, taking law in this last sense, by his own definition, Mr. Darwin does, nevertheless, continually bring forward certain "laws" as accounting for certain results. Thus, we have the laws of "Correlation of Growth, " [Footnote: Origin of Species, ed. 1872, p. 114.] "Inheritance limited to Males, " [Footnote: Descent of Man, vol. i. pp. 256, 257.] and a "Principle of Compensation. " [Footnote: Origin of Species, p. 117.] When Mr. Darwin, therefore, brings forward these laws as efficient causes, he not only tacitly admits the inadequacy of his theory to account for the phenomena in question, but he also endeavours to supply the defect by another cause, which, by his own definition, is no cause at all. And further, Mr. Darwin calls in the action of "unknown agencies. " [Footnote: Descent of Man, vol. i. p. 154.]

But it may be said, "Is not this the case with all sciences, at least in their earlier stages? Are there not frequently, or always, many phenomena which at first seem inexplicable, but which are gradually accounted for as knowledge increases? If, then, this is no objection in scientific pursuits generally, why should it be so here? " This reasoning would be perfectly valid if Darwinism were regarded simply as a scientific investigation. But it is under consideration now on very different rounds. Whatever Mr. Darwin's own views may be, the theory is brought forward by others, not as a mere interesting speculation, but as antagonistic to a record whose authority is attested by evidence of the very highest class. It claims to discredit that record, and to be received as a substitute for it. But that record, however it may be interpreted, does give us adequate causes for all that it professes to account for, in the will and operation of an Almighty Creator. The theory, therefore, which professes to supplant it, must at least stand upon an equal ground—it must give an adequate account of everything. There must be no unverified laws. To fall back upon such laws is in reality to fall back on the working of that very power whose operation is formally denied. [Footnote: See Foster's Essays, Essay i. Letter 5.]

The next point to be noticed is a great confusion between assumptions and proved facts. This is especially prominent in that part of his last work which is devoted to sexual selection. Thus, in one case it is taken for granted, that various characteristics of the males "serve only to allure or excite the female. " [Footnote: Descent of Man, vol. i. p. 258.] "Hence" (because brilliant colours of insects have probably not been acquired FOR THE PURPOSE of protection), "I am led to suppose that the females generally prefer, or are most excited by the more brilliant males. " [Footnote: Ibid. p. 399.] "Nevertheless, when we see many males pursuing the same female, we can hardly believe that the pairing is left to blind chance; that the female exerts no choice, and is not influenced by the gorgeous colours, or other ornaments with which the male alone is decorated" [Footnote: Descent of Man, vol. i p. 421.] Such sentences are of continual occurrence, and do duty in the argument as if they expressed ascertained facts. And not only this, but in the very part of the work which is devoted to establishing the adequacy of sexual selection to produce certain effects, that adequacy is assumed from the very beginning. Thus, we read, "That these characters are the result of sexual selection is clear, " [Footnote: Ibid. p. 258.] before we have got six pages into an argument which occupies a volume and a half. This is surely a strong instance of what is commonly called "begging the question. " Another instance of confusion of ideas is to be found in the assumption of design which occasionally occurs. Thus, we read, "In some other remarkable cases beauty has been gained for the sake of protection, through the imitation of other beautiful species. " [Footnote: Ibid. p. 393.] "From these considerations Mr. Bates inferred, that the butterflies which imitate the protected species, had acquired their present marvellously deceptive appearance through variation and natural selection, in order to be mistaken for the protected kinds. " [Footnote: Descent of Man, vol. i. p. 411.] In these cases there is an assumption of purpose and design, which, necessarily implies a designer, just as law, treated as an efficient cause, implies a law-giver. It may indeed be that this is only an inaccurate way of expressing something else; but then, such modes of expression are usually the result of a want of clear perception of the ideas to be expressed; and, in this case, such expressions must diminish the weight to be assigned to Mr. Darwin's judgment.

We come now to the consideration of the first of Mr. Darwin's supposed agencies—"Natural Selection, " or, "Survival of the fittest. " The results produced by this process must be ascribed to one of two causes: either they are the work of a Superintending Providence, watching over and directing every separate detail; or they are the result of pure chance and accident. There is nothing intermediate between these two causes. Natural law—apart from design and a designer—is, as we have seen, a nonentity—a mere expression of observed facts, for which it can give no account whatever. Mr. Darwin's argument is expressly directed to exclude the interference of a superintending Providence. Chance is the only cause which he can bring forward. The very first question, then, which arises is, What is there upon which chance may operate? What are the conditions from which the probabilities may be calculated? Mr. Darwin assumes, and no doubt correctly, that minute variations are continually taking place. But as these variations are the result of accident [Footnote: If they are not the result of accident, we again see design and need a designer.] they will take place in various directions; some of them will have a beneficial, some of them a noxious tendency. As, moreover, they are supposed to be very small at each step, the difference of advantage in the case of different individuals must be also very small, and will not be likely to produce any considerable difference in the chances of pairing. But in order that any variation may be perpetuated and increased, the pairing of similarly affected individuals is necessary. Parents, in which the variations took opposite directions, would probably have offspring of the normal type, the opposite variations neutralizing each other. And this must be repeated again and again; and with every repetition of the process required, the probabilities against it would rapidly increase. Thus, supposing that in the first generation the proportion of favourable conditions were such, that of those animals that paired there were four of each sex that had them to three that wanted them, the chances that any given pair were alike in possessing them would be represented by the product $4/7 \times 4/7$, or $16/49$. Hence, the chances would be rather more than two to one against it. In the next generation it would be $256/2401$, or more than eight to one, and so on. [Footnote: This is given merely as an illustration of the nature of the calculation. In any actual case the conditions would be infinitely more complex, but the calculation, if it could be made at all, must be made on this principle.]

But next, we have not to do with one series of changes only, but with a vast number of different series going on in different directions, if we are to have a large variety of animals produced from a common stock. All the probabilities against the separate variations must be combined, not by addition, but by multiplication, so that the probabilities against the production of all these separate forms become enormous.

Against all this improbability Mr. Darwin brings forward the supposed advantages which these variations give to their possessors. But here again a new element is introduced into the calculation. It is assumed, in the very statement of the question, that the process of adaptation has already taken place; the original stock must have been adapted to the circumstances under which they existed, or in their case the whole theory fails. If, then, a fresh adaptation is wanted, it must be because a change in external circumstances must have taken place. In order that a new variety may be established there must be a concurrence between the change of external circumstances and the change in the animals. Here we get a new, and a large factor for our multiplication.

This argument may be, perhaps, made clearer by an illustration. Mr. Darwin has written a very interesting book on the fertilization of orchids by means of insects. According to his view all insects are descended from one common type, and all orchids are also descended from one parent; but we meet with insects and orchids in pairs, each perfectly adapted to the other. We will suppose that a change takes place in a particular orchid, that the nectary recedes to a greater distance from the point to which the insect can penetrate, and so an advantage is given to those insects in which the haustellum is of a length above the average. This may have a slight tendency to increase the number of such insects; but then it will have an opposite tendency in the case of the orchid. It cannot, of course, be supposed that the variation, which is only partial in the insect, is universal in the plant. The unchanged insects will therefore be confined to the unchanged flowers, while the changed insects will be indifferent on the subject, as they will be able to reach the nectary in any case. Hence, an advantage will be given to the unchanged flower, which will be more likely to be fertilized, and the two lines of variation will move in opposite directions.

But next, the variation in the insects and the flowers must take place at the same time and the same place, or no result will follow to the insect, while the new variety of orchid must perish for want of an insect to fertilize it. It is this which makes the supposition of unlimited time almost useless, because just in proportion as the time is increased the probability of two independent events happening simultaneously is diminished.

But even supposing this difficulty out of the way, we meet with an immediate repetition of it. The insect derives an advantage from its increased haustellum, but what advantage does the plant derive from its retiring nectary? How does that help it in the "struggle of life? " But if it produces no beneficial result, the variation according to the theory must drop. Hence we should arrive at an insect suited for a new form of the flower, but no flower suited to the new form of the insect.

If, then, we reject the idea of superintendence and design, we have on the one hand an enormous antecedent improbability, while on the other hand we have only a very small power by which a direction may be given to the course of events, since by the hypothesis in any one generation the change, and consequently the superior advantage, is exceedingly small, and there is a strong tendency in related changes, as in the case of the orchid and insect, to move in opposite directions.

But next, in the varieties of animals with which we are acquainted, there is a certain connexion between the differences of independent organs, for which this theory does not help us to account. Thus, for instance, according to this theory the canine and the feline races are descended from a common ancestor. But there are several points of difference between a cat and a dog. There are the differences in the form of jaws, in the dentition; in the muscles by which the jaws are moved, and in the feet and claws. All animals of the cat tribe agree in all these respects, so do all animals of the dog tribe. We never find a cat's head combined with the feet of a dog. Why is this? Mr. Darwin attempts to account for it by his supposed law of "correlation of growth, " but, as has been already shown, any such law, being by Mr. Darwin's definition the observed sequence of events and nothing more, is utterly useless, when it is brought forward as a cause for those events. On this point the theory completely breaks down.

3. The theory does not account for any changes which are not immediately beneficial. [Footnote: In the "Origin of Species" (Ed. 1872) Mr. Darwin makes an admission which is virtually a giving-up of his whole theory. He says, "In many other cases modifications are probably the direct result of the laws of variation or of growth, independently of any good having been thus gained; but even such structures have often, as we may feel assured, been subsequently taken advantage of, " pp. 165, 166. Here, then, we have a preparation for future circumstances, which surely implies design.] If any rudimentary advance is made in the organism, if, for instance, the rudiments of a new bone, or joint, or organ of sense are developed, the nascent organ must, according to the hypothesis of minute changes, be useless in the first instance. Hence it would confer no advantage in the struggle of life; there would be no tendency towards its preservation and growth. This becomes a very important consideration, when certain important differences in animal structure and habits are to be accounted for. How, for instance, could the mammary glands be developed in oviparous creatures? Mr. Darwin regards them as originating in cutaneous glands, developed in the pouch of the marsupials. But his grounds for this statement are very meagre. To a great extent they rest on what an American Naturalist "believes he has seen; " and besides, the ornithorhyncus, which has no pouch, and which is lower in the scale of life than the marsupials, by Mr. Darwin's own admission (O. S., p. 190), possesses the glands. Mr. Mivart's question (Darwin, O. S., p. 189) is a very pertinent one.

Another point which this view fails to explain, is the determination of the line of development in particular directions at different periods. At one time it is most marked in fishes, at another in reptiles, at another in mammals. How is this to be accounted for?

4. The experience of cattle-breeders does not warrant the assumption that the principle of natural selection has more than a limited operation. No case has as yet been brought forward in which varieties have been produced which were not capable of interbreeding. Apart from their experience there is not a particle of evidence in favour of the assertion that races which cannot be made to breed together can be descended from a common stock. The unlimited application of this principle is therefore a pure assumption.

5. To this must be added the circumstance that no authenticated instance of variation by natural selection can be brought forward. It is true that this is not a very important argument, because our knowledge of those classes of animals in which natural selection could act is even now very incomplete; and our knowledge of their past history is still more limited, so that we are not in a condition to prove a negative. But in such a case as this the onus of proof should surely lie on the other side. It is for those who would assert the theory to bring forward positive proof of it. There is, however, one point in Mr. Darwin's view of domesticated animals which tells against his theory. The cat remains unchanged, because from its vagrant habits man has no control over its pairing [Footnote: Darwin's "Animals and Plants, " vol. ii. p. 236.]. Now considering the variety of conditions under which cats exist, here is surely a great opening for natural selection. But it has produced no results.

We come now to the theory of Sexual Selection, which is to account for those peculiarities and distinctions which can have no beneficial effect in the struggle of life, and which are accounted for on the supposition that they render their possessors more agreeable to the opposite sex, and so facilitate pairing, so that those animals which possess them in a remarkable degree would have the greatest chance of continuing their race. The case on which Mr. Darwin mainly rests his argument is that of birds, in which the males are frequently distinguished by exquisite colours and very graceful markings, and in which also the proceedings of the sexes can, in many cases, be more easily watched.

It is in maintaining this theory that Mr. Darwin has such frequent recourse to what may be called the "argumentum ad ignorantiam. "

"If such and such organs or ornaments were not designed for this or that particular object, then we do not know of what use they are. " [Footnote: For instance, Descent of Man, vol. ii. pp. 284. 399.] This maybe very true, but it proves nothing, unless we assume that we are or ought to be acquainted with, the use and object of everything in nature. And it involves another and a very wide question. There are certain tastes which seem to be inherent in our nature, and there are certain external objects which afford gratification to those tastes. Must we view this coincidence as merely accidental? or is it a part of the design of the world that it should minister not only to our needs, but also to our enjoyments? Mr. Darwin does not reject the idea of an Author and Designer of Nature, is he then prepared to assert that beauty did not form a part of the design as well as utility? [Footnote: In the "Origin of Species, " p 159, Mr. Darwin does seem to assert this; but he says in conclusion, "How the sense of beauty in its simplest form—that is, the reception of a peculiar kind of pleasure from certain colours, forms, and sounds—was first developed in the mind of man and of the lower animals is a very obscure subject, " p. 162. To Mr. Darwin, with his present views, it may well be obscure; but it presents no obscurity at all to those who believe that the universe in all its details was designed, and its formation superintended, by a loving Father, whose will was that it should not only supply the needs, but also minister to the enjoyment of all His creatures, nor to those who in every form of beauty, physical, intellectual, or moral, behold a far-off reflexion of the glory of the Invisible Creator.] If he is not prepared to assert this, he must admit the possibility that many things exist whose sole object is to minister to that sense of beauty which is probably possessed by other beings besides ourselves.

Mr. Darwin admits that many other causes, beside the supposed preference on the part of one sex for certain material adornments possessed by the other, influence the pairing of animals. In a very large number of cases the female is quite passive in the matter. The question is decided by a battle between the males, and the female seems, as a matter of course, to become the mate of the conqueror. In many other cases pairing seems to be the result of accident; the two sexes pair as they happen to meet each other. The great points on which Mr. Darwin rests his argument are that in some cases, on the approach of breeding-time, certain ornamental appendages become more highly developed or more brilliantly coloured, [Footnote: Descent of Man, vol. ii. p. 80.] and that in many cases the males, when courting the females, are observed to display their ornaments before them. [Footnote: Ibid. vol. ii. p. 86, et seq.] but then there are other facts, which Mr. Darwin. also notices, which detract more than he seems willing to allow, from the relevancy of these facts. The development of ornaments at breeding-time sometimes takes place in both sexes, indicating some latent connexion with the reproductive organs; thus the comb of the domestic hen becomes a bright red, as well as that of the cock. It would appear then that the object of the change is not to render the cock more attractive to the hens, for how could it serve the hens (if the choice lies with them) to be made more attractive to the cocks? Then again an old hen who is past laying, often assumes, to a considerable extent, the plumage of the cock. When these ornaments are the exclusive possession of the male, they are often displayed for other purposes than the gratification of the female. The possessors seem to be conscious of their beauty, and to take a pleasure in displaying it to any spectators.

Very great beauty and brilliancy of colour is often found in cases in which it can have nothing whatever to do with the relation between the sexes. Thus, a vast number of caterpillars are remarkable for their beauty; but in their immature state it can have no relation to sexual selection; and if it may, or rather must, have a different object in one case, what ground have we for assuming that it may not have a different object in the other?

Again, we are not in a position to form any opinion as to the causes which really influence the pairing of animals when choice is exercised. We have no certain knowledge upon the important question whether the ideal of beauty, if possessed by the lower animals at all, is in all, or even in many cases, in accordance with our own. We, for instance, admire a male humming-bird; what certainty have we that he is equally beautiful in the eyes of his mate? In cases where we have reason to believe that deliberate selection has taken place, we do not know that that selection was influenced by only one condition—that of beauty. There may have been a thousand causes at work of which we know nothing. Mr. Darwin brings forward an instance in which the owner of a number of peahens wished them to breed with a peacock of a particular variety, while they showed a deliberate preference for another bird; and he supposes that their preference was decided by the plumage. But there might have been another cause—at least the circumstances as related by him seem to suggest it—which would give a very different turn to the affair. The favoured peacock, spoken of as "old, " [Footnote: Descent of Man, vol. ii. p. 119.] was probably an old friend of the hens, while his unsuccessful rival seems to have been a new introduction. The preference shown by the hens would in this case be fully accounted for, without supposing them to have exhibited any choice in the matter of plumage.

Then there are a vast number of peculiarities which are certainly not ornamental in our eyes, but which are confined to the male sex. They are, so far as we can tell, of no service whatever in the struggle of life. With reference to these Mr. Darwin's argument seems to be this, —"They can serve no other purpose with which we are acquainted, therefore they must be attractive to the female—therefore they must be acquired by sexual selection. " Such arguments as these cannot carry much weight. [Footnote: Descent of Man, vol ii p 284.]

On the whole, we can hardly come to any other conclusion than that the theory of sexual selection is not proved. In many cases it is known that such selection is not the result of choice; in other cases, where choice seems probable, we have no ground for believing that external appearance is the sole ground of that choice. It may exercise some influence, but that is all. Even if admitted, there are many things which cannot be accounted for by it without very extravagant assumptions. It cannot then be admitted as covering the large classes of phenomena left unaccounted for by the theory of natural selection.

So far as the lower animals are concerned, the results to which an examination of Mr. Darwin's views has led us may be summed up in the following propositions: —

1. That the two causes, natural and sexual selection, have probably exercised some influence in the modification of animal forms; but that the laws of probability preclude our entertaining the belief that these causes can have had, by themselves, and apart from a superintending power, anything beyond a very limited operation.

2. That in cases where there have been related changes in different parts of the same organism, or in different organisms, the inadequacy of these two causes is virtually admitted by the introduction of certain supposed laws; and that these laws, being defined by Mr. Darwin to be no more than "the ascertained sequence of events, " cannot be regarded as efficient causes, and so cannot supply the defect.

3. That there are particular points in the chain of life, in which the transition from one form to another is so great, and so incapable of graduation, that it is impossible to suppose that these two causes can have been adequate to produce it. Of this a notable instance is to be found in the transition from oviparous animals to the mammalia.

We come now to the consideration of the origin of man, which Mr. Darwin, in his last work, ascribes also to natural and sexual selection. His view is, that man is descended from some family of anthropomorphous apes, and that all those enormous differences which, as he admits, exist between the highest ape and the most degraded member of the human race, are differences of degree only, and not of kind; that all our intellectual wealth, and all our moral laws, are simply the development of faculties and ideas which were possessed in a ruder form by the creatures from whom man is descended.

So far as man's physical constitution is concerned, there is undoubtedly something to be said in favour of this view. For man's bodily frame is composed of the same elements, and moulded upon the same general plan as that of the higher apes, and, what is still more remarkable, it retains, in a rudimentary form, certain muscles and organs which are fully developed and answer important purposes in many of the quadrumana. Of these the tail is a remarkable instance. But when the differences between the physical peculiarities of man, and those of his supposed progenitors are examined, the theory of natural selection collapses entirely, for the development has taken the form which would be most disadvantageous in the struggle of life. This is very clearly put by the Duke of Argyll. [Footnote: "Recent Speculations on Primeval Man, " in Good Words, April, 1868.]

"The unclothed and unprotected condition of the human body, its comparative slowness of foot; the absence of teeth adapted for prehension or for defence; the same want of power for similar purposes in the hands and fingers; the bluntness of the sense of smell, so as to render it useless for the detection of prey which is concealed; —all these are features which stand in fixed and harmonious relation to the mental powers of man. But, apart from these, they would place him at an immense disadvantage in the struggle for existence. This, therefore, is not the direction in which the blind forces of selection could ever work .... Man must have had human proportions of mind before he could afford to lose bestial proportions of body. "

But it is in the intellectual and spiritual part of man's nature that the greatest difficulty in the way of the application of these theories arises. The strongest argument of all against them is one which is incapable of proof, since it arises not from facts around us, but from our own self-consciousness—our realization of our own powers—and so, to each individual man it must vary in apparent strength, in proportion as he realizes what he is, and what it is in his power to become. The very outcry that has been raised against Mr. Darwin's proposition is a proof of this. The theory of the descent of man, as he propounds it, was felt to be an outrage upon the universal instincts of humanity. But, because this objection rests upon such a foundation, it is incapable of being duly weighed and investigated as an argument, and we proceed therefore to such considerations as are within our reach.

First of all it is desirable to dispose of one of the stock arguments in favour of the theory. That argument is, that the difference between the lowest type of savage and the highest type of civilized man—between a Fuegian or an Australian on the one hand, and a Newton, a Shakspeare, or a Humboldt, on the other, —is quite as great as that between the higher forms of ape and the lowest forms of humanity. But in this argument there is a fatal confusion of ideas. The capacity for acquisition is confounded with the opportunity for acquisition. That the savage is in possession of but very few ideas does not prove that he is incapable of more; it may equally well arise from the fact that he had had no opportunity of acquiring more. The only way to test the question is by putting a savagoe from his earliest infancy, under the same favourable circumstances as the child of civilisation. Whenever this experiment has been tried, and our missionaries have had many opportunities of trying it, the difference has either not appeared at all, or has proved to be very trifling. Mr. Darwin himself seems to have been very much surprised at what he saw in some natives of Terra del Fuego, who were for a time his companions on board the "Beagle. " "The Fuegians rank amongst the lowest barbarians, but I was continually struck with surprise how closely the three natives on board H. M.S. 'Beagle, ' who had lived some years in England, and could talk a little English, resembled us in disposition, and in most of our mental faculties. " [Footnote: Descent of Man, vol. i. p 34] And these Fuegians had not been educated from their infancy, they had only come to England later in life, and were thus under an incalculable disadvantage. Had they been heirs to such an intellectual inheritance as fell to the lot of Mr. Darwin, there is nothing extravagant in the supposition that they might have proved themselves equal to him in the ability to make use of it. The comparison then proves to be quite illusory; but it draws our attention to a fact which is of very high importance in our investigation of the difference between man and all other animals. Man alone seems to be capable of laying up what may be termed an external store of intellectual wealth. Other animals in the state of nature make, so far as we know, no intellectual advances. The bee constructs its cell, the bird builds its nest precisely as its progenitors did in the earliest dawn of history. There is a possibility that some advance, though a very small one, may be made by animals brought under the control of man. It is said, for instance, that a young pointer dog will sometimes point at game without any training. But in this case the acquired knowledge is congenital, and is therefore to be regarded as a development brought about by superintended selection. But with man none of the acquired knowledge is innate. It is a treasure entirely external to himself until he has appropriated it by study of some kind or other. There is no reason to believe that any advance in intellectual power has been made by man, in his collective capacity, since his first appearance on earth. Various individuals have varying powers, but these differences are no result of development, since they may often be found among members of the same family, who have been subjected to the same discipline, and enjoyed the same educational advantages. It follows that the gulf between the ape and the lowest type of humanity is almost if not quite as great as between the ape and the highest type. The savage does not in any way help to bridge over that gulf.

But it is said that the moral and intellectual faculties which man possesses, and which he looks upon as the great badge of his superiority, are in truth only different in degree and not in kind from those possessed by the lower animals. But the grounds on which this assertion is based are wonderful in their tenuity. Dogs are possessed of self-consciousness because they sometimes emit sounds in their sleep from which it is concluded that they dream. [Footnote: Descent of Man, vol. i. p. 62.] "Can we feel sure that an old dog, with an excellent memory, and some power of imagination, as shown by his dreams, never reflects on his past pleasures in the chace? And this would be a form of self- consciousness. " Our duty to our neighbour is entirely the result of "social instinct, " [Footnote: Descent of Man, vol. i. pp. 70- 106.] and our duty to our God the development of a belief which has its origin in dreams. [Footnote: Ibid, p. 66.]

It is impossible for us satisfactorily to meet these assertions with a direct negative, [Footnote: There are some who think that this statement may be directly refuted. Their views will be found in the QUARTERLY REVEIW, July, 1871.] for this simple reason, that we have no means whatever of knowing what ideas are present in the minds of the lower animals, or even what communications pass between them. For anything we can tell to the contrary, the bark of a dog may be as articulate to his fellow-dogs as our speech is to our fellow-men, while on the other hand to the dog our speech may be as inarticulate as his bark is to us. But our total ignorance of the mental state of animals which have been the companions of man from the very earliest ages, our utter inability to hold any conversation with them, is in itself a proof of the wide gulf that separates them from us. Put two men of the most widely separated races on a desert isle together, and a very little time will elapse before they are able to hold some communication with each other. If then the difference between man and the lower animals were a difference of the same kind as that between the civilized man and the savage, though greater in degree, surely in so many thousand years something might have been done to open a way for intellectual communication; some development of the faculties of the lower creatures would have been perceived, some means of interchanging ideas would have been discovered. If Mr. Darwin had had for his companions on board the "Beagle, " instead of three Fuegians, as many Gorillas or Chimpanzees, would he, at the end of the voyage, have been able to report any approximation, at all to European mental characteristics, or even to those of the lowest savage? But if the difference be only one of degree, some approximation ought to have taken place.

As then we can have no direct knowledge of the moral and intellectual powers of animals, we can only judge of them from their actions, and other external signs. One great mark of difference has already been noticed. Man has, other animals have not, the power of laying up an external treasure of intellectual acquirements. Then there are certain arts which seem to be indispensable to man in his lowest state—no savage is so low that he is utterly destitute of them—no animal makes any pretence to them. Such are the designing, construction, and use of tools. Mr. Darwin asserts that in certain cases—very rare ones—apes have been known to use stones to break open nuts; but the mere use of a stone is a very different thing from the conception and deliberate formation of a tool, however rude. Then there is the kindling of fire, and the use of it for the purpose of cooking; and lastly, the preparation and the wearing of clothes. The tools or the clothes may be of the rudest kind, the tools may be formed from a flint, and the clothes from bark or skin, but in the preparation of each there are signs of intellectual power, of which we find no indications whatever in the lower animals.

Another important difference between man and all other animals lies in the fact, that whatever an animal does it does perfectly from the first, but it makes no improvements. A bird's first nest is perfect. With man the case is the reverse, it is only by many trials, many failures, that he attains to skill in any operation, but then he goes forward. Arts improve from generation to generation. This seems to show that the faculties of man differ from those of animals in kind, and not in degree only.

The question also arises, if man has been produced from an anthropomorphous ape by a process of natural development, how is it that the same process has not gone on in other lines? The dog, the horse, and the elephant are at least equal in intelligence and sagacity to the highest known apes. Such a development from them cannot have proceeded through the line of the apes. If these different orders are at all connected it must be through some remote common ancestor. Why then has this development come to an abrupt termination in some cases and not in all? It may indeed be said that the dog and the horse are indebted for their intelligence to the inherited results of long intercourse with man, but this cannot be the case with the elephant, which is never known to breed in captivity. Nor is there any reason to believe that the present intelligence of the elephant is recently developed. Why then has it been arrested in its course?

Whether or not we assume the theory of development to be wholly or partially correct in reference to the lower animals, we must admit that it is true of man, but in a sense totally different from that which Mr. Darwin suggests. The development of which he is the advocate is a development of race, in which the advance made by each individual generation is exceedingly small, while the difference in remote generations, the accumulated advance of successive generations, is great. In man, on the contrary, there is no reason whatever to believe that there has been any advance at all in the race from the very earliest periods—that either in physical power or intellectual ability the present generation of men, taken as a whole, are in any way superior to their most remote ancestors. The development of which man is especially capable is the development of the individual, that development being not physical, but intellectual and moral, and being in a great degree dependent on the will and perseverance of the individual, and very little on external circumstances. The result of these individual developments has been the accumulation of a vast fund of wealth, useful arts, sciences, literature, which form the common possession of the whole race, but do not necessarily imply the slightest advance in any particular individual—that advance being dependent, not on the possession of those treasures, but on the use made of them. In the case of man then development does certainly exist, but it takes a line totally distinct from that which Mr. Darwin advocates, and thus forms another broad line of demarcation between man and the most advanced of the lower animals.

It appears then that the faculties of man differ generically from those of the animals. A new order of things seems to have commenced with the appearance of man on the earth—an order in which the highest place was to be maintained by intellectual instead of physical power. No mere process of evolution then will account for man's origin. His physical nature may have been formed in that way; but we cannot believe that his intellectual and moral nature were developed from any lower creatures. Only some special Creative interference can account for his existence.

So far then as it tends to negative the continued operation of the Creator, the theory of evolution is untenable. Like that of Laplace, it fails to give an adequate cause for existing phenomena. But it seems probable, as will be seen in the next chapter, that both theories have in them much of truth. They cannot point out the cause of the universe, but they may give us a more or less accurate view of the manner in which that cause operated. The facts brought forward by geologists have been shown not to be incompatible with interpretations which the Mosaic Record readily admits, though they conflict with existing notions upon certain points. In no one then of the three sciences which have been supposed to be specially antagonistic to that record, is there anything to be found which can be maintained as a reasonable ground for doubting that that record is, what it has always been held to be by the Church, a direct Revelation from the Creator.

CHAPTER V.

SCIENCE A HELP TO INTERPRETATION.

It is now clear that there is nothing in the Mosaic Record itself, which is contradicted by any scientific discovery, and that all the alleged difficulties arise either from interpretations prematurely adopted, or from theories which, when carefully examined, are found to be defective, but which may nevertheless contain in them a large element of truth. But if scientific discoveries are available for the refutation of erroneous interpretations, the probability is that when rightly understood they will help us to arrive at the true meaning, since the Works of God are, beyond all other things, likely to throw light on that portion of His Word in which those Works are described. Nor are the theories to be passed over—the greater the amount of truth which they embody the greater will be the likelihood that they will receive help from, as well as throw light upon, such a record; and thus we shall have additional evidence that the Word, the Work, and the Intellect, which has scrutinized and interpreted the Work, are all derived from the same source. We proceed, therefore, to inquire whether these facts and theories do in any way elucidate the concise statements of Scripture, so that we may be enabled to arrive at a somewhat clearer idea of the meaning of this most ancient document, and be enabled to entertain somewhat more distinct views of the manner in which the Divine Architect saw fit to accomplish His Work.

In pursuing this investigation two points must be carefully kept in mind; the first is the distinction between theory and conjecture on the one hand, and well ascertained facts on the other. We shall have much to do with theory, and with conjectural interpretations of observed facts. These can never stand on the same footing as the facts themselves, but can only be regarded as invested with greater or less probability. If it is found that these theories do explain many observed facts, that they harmonize with, and as it were dovetail into any proposed interpretation of which the words of Moses are capable; and still more if that interpretation actually completes the defective points of the theories, and supplies an adequate cause for facts hitherto inexplicable—then the presumption is a very strong one that the interpretation thus supported is at all events an approximation to the true one.

The second point to be carefully kept in mind is the very imperfect state of scientific knowledge even at the present time. As far as the matter in hand is concerned, the facts which are ascertained beyond all possibility of doubt, are very few. New means of investigation have very recently been discovered, and as a consequence new sources of information have been pointed out, new fields of research have been laid open. Twenty years ago the spectroscope was a thing undreamt of—now astronomers reckon it as of equal value with the telescope, while chemists find it indispensable to their researches. Who shall say that the next twenty years may not witness some invention of equal importance, which shall throw upon us a fresh flood of light from some unexpected quarter? If then the principle which has hitherto been maintained is correct, that all our difficulties arise from interpretations based upon insufficient knowledge, but maintained as if of equal authority with the record itself, there is a great danger lest after a time the same difficulty should recur—that the discovery of fresh facts may discredit interpretations based upon our present knowledge. Any interpretation therefore to which we may be led by the scientific views at present entertained, must be regarded as only provisional and tentative, liable at any time to be either confirmed, amended, or rejected, as fresh discoveries may be made.

Before we enter upon a detailed examination of the records of the several days, there are two preliminary points to which attention must be directed. We shall have to make frequent reference to "law. " It will be well that the sense in which the term is used should be made clear. The account of the First Day's Work will lead to the recent theory of the Correlation of Forces. As this is probably a new subject to many, some previous explanation of it will be necessary.

SECTION 1. OF LAW. [Footnote: This subject is fully treated in the Duke of Argyll's "Reign of Law. "]

Law, in its original and proper sense, is the expression to an inferior of the will of a superior, which the inferior has it in his power to obey or to resist, but resistance to which entails a penalty more or less severe, in proportion to the moral turpitude, or the injurious consequences of the act of disobedience. In this its strict sense the law can only exist in connection with beings possessed of reason to understand it, of power to obey it, and of free will to determine whether they will obey it or not. When these three conditions are absent law can have no existence. But the result of perfect law, perfectly obeyed, would be perfect order. Hence the observation of perfect order leads, by a reversed process, to the supposition of some law of which that order is the result. Hence arose in the first instance the term "natural laws, " or "laws of nature. " Events were found to follow each other in a uniform way, and this uniformity was thus sought to be accounted for. Probably in the minds of those by whom the word was thus applied in the first instance Nature was not the mere abstraction it is now, but an unseen power—Deity or subordinate to Deity— working consciously and with design.

[Footnote: Mr. Darwin, especially in the "Origin of Species, " seems continually to betray the existence of this feeling in his own mind. Though he from time to time reminds us that by Nature he means nothing but the aggregate of sequences of events, or laws, he yet frequently speaks of Nature in a way which is applicable only to an intelligent worker.]

But this feeling has disappeared, and now we are told that natural law is "the observed sequence of events. " In this case, then, the true meaning of the word is entirely lost—it is no longer possible to speak of law as the cause of any event.

But the old sense in which the word was applied to natural phenomena had in it far more of truth than the modern one. It was the imperfect expression of the great truth that God is a God of order—that there is a uniform procedure in His works, because in Him there is no change, no caprice. And it is of great importance to us that we should realize this truth, because we are dependent upon the laws of nature every moment of our lives. Every conscious act is performed under the conviction that the natural forces which that act calls forth will operate in a certain prescribed manner. But this conviction, though it restricts us to the limits of the possible, does not further impede the freedom of our will. To a certain extent we can choose what action we will perform, what forces we will call forth for that purpose, and what direction we will give them. Sometimes we can arrange our forces so that they will continue to act for a considerable time without any intervention from us; in other cases continued interference is necessary. But in all these cases there is no interruption of the law by which the working of these forces is regulated. We have then a limited control over these forces, and yet they are unchangeable in themselves, and in their mode of action.

When, however, we strive to ascend from our own works to those of God, we can no longer regard these forces as absolutely unchangeable. If they are practically so, it is because it is His Will that they should be so. It is this Will then which has its expression in the so-called laws of nature. The term now assumes a sense akin to, though not identical with, its original ethical sense. It is no longer a rule imposed by a superior on an inferior, but the rule by which the Supreme Being sees fit to order His own Work. While however we admit the possibility of law of this kind being changed, we have no reason to believe that in the universe with which we have to do any such change has ever taken place. But this does not preclude the possibility of Divine interference in the processes either of Creation or of Providence. New forces may from time to time be supplied, new directions may be given to existing forces, without any variation in the laws by which the action of those forces is regulated.

And if we believe that Creation was a progressive act, it is rather probable than otherwise that such interferences should take place. For a long period perhaps the uniformity of the work might lead us to forget the Being who was working; but times would arrive when definite stages of the work were accomplished, when higher developments of being were rendered possible, and in the introduction of those higher developments a something would be seen which could not be the result of the processes with which we had already become acquainted. Such interference would not in any way justify the supposition that the designs of the Author of Nature were changed, or that His original plan had proved defective. The more natural inference would be that they were a part of the plan from the first, but that the time for them was not then come.

It will be seen in the sequel that in all probability many of the special acts of Creation, mentioned in the Mosaic Record, are interferences of this kind; that for long periods of time matters advanced in a uniform manner; that the sequence of events was such as our own experience would lead us to anticipate; but that these periods were separated from one another by the introduction of new forces and new results. Of the former we may speak then as carried on under the operation of natural laws; the other may be described as special interferences not antagonistic, but supplementary, to natural laws, and forming part of the original design.

SECTION 2. THE CORRELATION OF FORCES.

[Footnote: For fuller information on this subject, Grove's "Correlation of the Physical Forces, " or Tyndall's "Lectures on Heat considered as a Mode of Motion, " may be consulted.]

It has long been known that heat and light are closely connected together. The accumulation of a certain amount of heat is always accompanied by the appearance of light. But when it was found that the light could be separated from the heat by various means, it seemed possible that the two phenomena were simply associated. It is now, however, ascertained that light and heat are identical in their nature, and that a vast number of other phenomena— electricity, galvanism, magnetism, chemical action, and gravitation, as well as light and heat, are different manifestations of one and the same thing, which is called force or energy. In a great number of cases it is possible for us, by the use of appropriate means and apparatus, to transform these manifestations, so as to make the same force assume a variety of forms. Thus motion suddenly arrested becomes heat. A rifle-ball when it strikes the target becomes very hot. The heat produced by the concussion against an iron shield is found sufficient to ignite the powder in some of the newly invented projectiles. The best illustration, however, is to be obtained from galvanism. By means of the Voltaic battery we set free a certain amount of force, and we can employ it at pleasure to produce an intense light in the electric lamp, or to melt metals which resist the greatest heat of our furnaces; it will convert a bar of iron into a magnet, or decompose water into its constituents, oxygen and hydrogen, or separate a metal from its combination with oxygen. But in all these processes no new force is produced—the force set free is unchangeable in itself, and we cannot increase its amount. Owing to the imperfection of our instruments and our skill a part of it will always escape from our control, and be lost to us, but not destroyed. When, however, due allowance is made for this loss, the results produced are always in exact proportion to the amount of force originally set free. Thus, if we employ it to decompose water, the amount of water decomposed always bears an exact proportion to the amount of metal which has been oxidized in the cells of the battery.

This force pervades everything which comes within the cognizance of our senses. It exists in what are termed the elementary substances of which the crust of the earth is composed. A certain amount of it seems to be required to maintain them in the forms in which we know them; for in many cases, when two of them are made to combine, a certain amount of force is set free, which commonly makes its appearance as heat. This seems to indicate that a less amount of force suffices to maintain the compound body than was requisite for its separate elements. Thus, when oxygen and hydrogen are combined to form water intense heat is produced. If we wish to dissolve the union, and restore the oxygen and hydrogen to a gaseous state, we must restore the force which has been lost. This, however, must be done by means of electricity, as heat produces a different change—converting the water into vapour, but not dissolving the union between its elements.

Force, in the shape of heat, determines the condition in which all inorganic bodies exist. In most cases we can make any given element assume the form of a solid, a fluid, or a vapour, by the addition or subtraction of heat. Thus if a pound of ice at 32 degrees be exposed to heat, it will gradually melt—but the water produced will remain unchanged in temperature till the last particle of ice is melted—then it will begin to rise in temperature; and, if the supply of heat be uniform, it will reach a temperature of 172 degrees in exactly the same time as was occupied in melting the ice. Thus then the force which was applied to the ice as heat passes into some other form so long as the ice is being melted—it is no longer perceptible by the senses—we only see its effect in the change from the solid to the fluid form. And this result is brought about by a definite quantity of force. Each of the inorganic materials of which the crust of the earth is composed seems thus to require in its composition a definite amount of force.

The life of vegetables is developed in the formation of fresh compounds of inorganic matter and force. No vegetable can thrive without sunlight, either direct or diffused. This supplies the force which the plant combines with carbon, hydrogen, and other elements to form woody fibre, starch, oils, and other vegetable products. When we kindle a fire, we dissolve the union which has thus been formed—the carbon and hydrogen enter into simpler combinations which require less force to maintain them, and the superfluous force supplies us with light and heat.

The life of animals is developed by a process exactly the reverse of vegetable life. It is maintained by the destruction of the compounds which the vegetable had formed. These compounds are taken into the body as food, and after undergoing certain modifications and arrangements are finally decomposed. Of the force thus set free a part makes its appearance as heat, maintaining an even temperature in the body, and another part supplies the power by virtue of which the muscles, &c., act. No manifestation of animal life is possible except by force thus set free. It seems all but certain that we cannot think a single thought without the decomposition of an equivalent amount of the brain. It must not, however, be concluded that force and life are identical. Force seems to be only the instrument of which the higher principle of life makes use in its manifestations.

Force then pervades the whole universe so far as it is cognizable by our senses. But we cannot conceive of force as acting, without at the same time conceiving of something on which that force acts. That something, whatever it may be, we designate "matter. " We have not the slightest idea of what matter really is—no man has ever yet succeeded in separating it from its combination with force. Even if success were possible, which seems very improbable, it is not likely that matter by itself would be discernible by any of our senses. We know that two of them, sight and hearing, enable us to perceive certain kinds of motion, i. e. manifestations of force, and this is in all probability the case with the rest of them. The existence of matter then is not known by scientific proof but by inference. Our belief in it arises from something in the constitution of our minds which makes it a necessary inference.

There is one more point in reference to force which must be noticed. It is indestructible, but it is capable of what is termed "degradation. " It may exist in various intensities and quantities, and a small quantity of force of a higher intensity may be changed into a larger quantity of force at a lower intensity. In the instance above given of the union of oxygen and hydrogen, heat is given out, but heat does not suffice to dissolve that union. The force must be supplied in the more intense form of Voltaic Electricity. But to reverse this process seems impossible for us. As, however, this is clearly explained in a previous volume of this series, [Footnote: Can we Believe in Miracles? p. 152.] it is not necessary to dwell upon it at length.

We may conclude then that the whole material universe is built up of matter and force in various combinations, but we can form no conception of what these two things are in themselves; they are only known to us by the effects produced by their union in various proportions.

SECTION 3. THE BEGINNING.

"In the beginning God created the heaven and the earth.

"And the earth was desolate and void, and darkness upon the face of the deep. "

These words carry us back to a time indefinitely remote. Eternity and Infinity are ideas which we cannot grasp, and yet we cannot avoid them. If we stretch our imagination to conceive of the most distant possible period of time—the farthest point of space— still we feel that there must have been something before the one, that there must be something beyond the other; and yet we cannot conceive of that which has no beginning, or no boundaries. The first verse marks out for us as it were a definite portion of this limitless ocean. "In the beginning, " is the point from which time begins to run—"the heavens and the earth, " the visible universe beyond which our investigations cannot extend. Whether other manifestations of God have taken place in Eternity, or other systems of worlds now exist in infinity, we are not told.

The heavens and the earth then are to be considered as comprising the visible universe, sun, moon, and stars, and their concomitants, which the eye surveys, or which scientific research brings to our knowledge. All are comprehended in this one group by Moses, and recent spectroscopic investigations teach us that one general character pervades the whole. Every star whose light is powerful enough to be analyzed, is now known to comprehend in its materials a greater or less number of those elementary substances of which the earth and the sun are composed. Whether any of these worlds were called into perfect existence at once, or whether they all passed through various stages of development, we are not told, that in some of them the process of development is only commencing, while in others various stages of it are in progress, is, as will be seen presently, highly probable. But the narrative takes no farther notice of anything beyond our own group of worlds, and proceeds to describe the condition of the earth (probably including the whole solar system) at the time at which it commences. Its words imply such a state of things as corresponds to what has been said in the preceding section of matter, apart from force. No better words could probably have been chosen for the purpose. The only word which seems to convey any definite idea is in the following clause, where water is mentioned. Until force was in operation water could not exist. Probably St. Augustine's interpretation is the correct one—the confused mass is called alternately earth and water, because though it was as yet neither one thing nor the other, it contained the elements of both. And the word "water" expressed its plastic character. ("De Genesi ad Literam" Liber Imperfectus, Section 13, 14.)

One other important point in these words is, that they negative the eternal existence of matter. The second verse describes it as existing, because it had been called into existence at the bidding of an Almighty Creator, as described in the first verse.

SECTION 4. THE FIRST DAY.

"And the Spirit of God (was) brooding upon the face of the water.

"And God said, 'Let light be' and light was.

"And God saw the light that it was good, and God divided the light from the darkness.

"And God called the light Day, and the darkness He called Night.

"And there was evening and there was morning, one day. "

The first clause seems to belong rather to the period of action than to the precedent indefinite period of chaos, and may therefore be taken as marking the transition from the "beginning" to the first day, better than as belonging to that beginning itself. The Jewish interpretation of the clause is untenable in the light of the doctrine of the Correlation of the Physical Forces. Till force was evolved there could be neither air nor motion, and so no wind. The words of course bear on their face an assertion of the action of the eternal Spirit in the work of Creation; but when we examine the position which they occupy, it seems highly probable that they have beyond this a much more definite signification. In them a sort of localized action is ascribed to the Spirit—a something very different from the idea conveyed by the often-repeated phrase, "And God said. " What that something may be it is hard for us to conceive, harder still to express, but the following considerations may perhaps throw some glimmering of light upon the matter: —

1. There must be some point in which the Creator comes into contact, as it were, with His creature—a point at which His Will first clothes itself in the form of a physical fact—the point to which all second causes lead up, and at which they lose themselves in the one first cause, the Will of God. Now this is what all systems of philosophy require as their starting-point, but it is entirely out of their unaided reach. But these words supply that indispensable desideratum.

2. These words come in immediate connexion with the evolution of light. Light is throughout the Bible intimately connected with the Deity. It is His chosen emblem. "God is light. " It is His abode. "He dwelleth in the light inaccessible. " It is the symbol of His presence, and the means by which Creation is quickened. "In Him was life; and the life was the light of men. "

3. Light, as we now know, is only one form of the force by which the universe is upheld. But the phenomena of light lead us to infer the existence of what we call Ether, which is supposed to be a perfectly elastic fluid, imponderable, and in fact exempt from almost all the conditions to which matter, as we know it, is subject, except that POSSIBLY it offers resistance to bodies moving in it. [Footnote: Encke's comet shows signs of retardation, as if moving in a resisting medium; but it is possible that that resistance may not arise from the ether, but from the nebulous envelope of the sun.] This fluid must pervade the whole universe, since it brings to us the light of the most distant star or nebula. As it is the medium through which light is conveyed, and as light is now known to be identified with force of all kinds, it seems by no means improbable that it is the medium through which all force acts.

These words, then, seem to suggest the idea that the brooding of the Spirit may have some connexion with the formation of that ether which is indispensable to the manifestation of light, and probably to the operations of all force; and that, if so, the ether may also be the point at which, and the medium through which, Spirit acts upon Matter. On the one hand, the facts that force, as used, is constantly in process of degradation, and that it is also constantly poured forth into space from the Sun and Planets in the shape of heat, and so lost to our system, seem to indicate that fresh supplies of it are continually needed; while, on the other hand, the supply of that need seems to be implied in the words, "By Him all things consist. " "Upholding all things by the word of His Power. "

If this be so, we have a point up to which natural laws may possibly be traced, but at which they merge in the action of the Will of God, which is beyond our investigation. Here, then, is a solution of that great difficulty, which those who are most familiar with the laws of nature have felt in reconciling the existence of those laws with a particular Providence and with the efficacy of Prayer, since we have here the point at which all forces and all laws begin to act, and at which, therefore, the amount of the force, and the direction of its action, are capable of unlimited modification, without any alteration of, or interference with, the laws by which that action is regulated, and consequently without the danger of introducing confusion into the Universe.

"And God said, 'Let light be' and light was. " It has already been pointed out that these words differ from those used in describing any other creative act. They are the only ones which seem to imply an instantaneous fulfilment of the command. Another matter which has long since been observed, is their exact harmony with what science teaches us respecting the nature of light. Light is not a material substance, but a "mode of motion. " It consists of very small undulations propagated with inconceivable velocity. Hence of it, and of it alone, it could not be correctly said that it was created. To say that God made light would be inexact. The words which are used exactly suit the circumstances of the case. But the discovery of the correlation of forces has given to these words a much more extended significance, while at the same time it furnishes a satisfactory reason for their occurrence at this particular point. So long as they were supposed to refer to light simply, they seemed out of place. Light was not apparently needed till there were organisms to whose existence it was essential. But we now know that to call forth light, was to call force in all its modifications into action. It has been seen that matter and force are the two elements out of which everything that is discernible by our senses is built up. The formation of matter has already been described in the original act of creation. But till force also was evolved, matter must of necessity remain in that chaotic state to which verse 2 refers. To matter is now added that which was required to enable the progressive work of Creation to be carried on. The first result of this would probably be that the force of gravitation would begin to act, while, from what the telescope reveals to us, we may conjecture, that at the same time the whole incoherent mass would be permeated with light and heat, and some, at all events, of those elementary substances with which chemistry makes us acquainted would be developed, and the whole mass, acted upon by the mutual attraction of its several particles, would begin to move towards, and accumulate about its centre of gravity.

It has been shown that Laplace's Nebular Hypothesis, when substituted for the action of a Creator, broke down in three important points. Of these the first two were, that it failed to give any account of the origin of matter, and of the first commencement of the action of Gravitation. These two defects are completely supplied by the first three verses of Genesis. We may probably see in the "Great Nebula" in Orion an illustration of the condition of the solar system when light first made its appearance. It is very probable that that nebula has only very recently become visible. Galileo examined Orion very carefully with his newly invented telescope, but makes no mention of it. [Footnote: Webb's Celestial Objects, p. 255, note.] At present it is visible to the unaided eye even in England, where the atmospheric conditions and its low altitude are alike unfavourable. In Italy, where the atmosphere is remarkably pure, and the meridian altitude is greater by 7 1/2 degrees, it must be a conspicuous object, and had it been so at the time when Galileo was observing the constellation, it could hardly have failed to attract his attention. It was, however, noticed in 1618. It is a vast, shapeless mass, having its boundaries in some parts tolerably well defined, while in other directions it fades away imperceptibly; its light is very faint, and when examined by the spectroscope is found to proceed from a gaseous source. Professor Secchi has traced it through an extent of 5 degrees. When it is remembered that at such a distance the semi-diameter of the earth's orbit subtends an angle less than 1 inch, some idea of the enormous extent of this mass of gas may be formed. Drawings of it have been made from time to time by our most distinguished astronomers, which are found to differ considerably. Great allowance must, of course, be made for differences in the telescopic power employed, and in the visual powers of the several observers, but the differences in the drawings seem too great to be explained by those sources of inaccuracy alone, and actual change in the nebula is therefore strongly suspected. Another nebula of similar character, in which changes are suspected, is that which surrounds the star A in the constellation Argo. This is being very carefully watched through the great telescope recently erected at Melbourne, and from the observations made there, it is probable that fresh light may soon be thrown on the subject.

The next act recorded is, that "God divided the light from the darkness. " This is one of those passages which we are very apt to pass over as unimportant, without giving ourselves any trouble to ascertain what they mean, or asking if they may not give valuable information, or supply some important hints. It is evident, however, that in these words some act of the Creator is implied, but when we inquire what that act was, the answer does not lie immediately on the surface. Darkness is simply the absence of light. It cannot therefore be said that God divided the light from the darkness in the same sense in which it is said that "a shepherd divideth his sheep from the goats". Between light and darkness that division exists in the very nature of things, and it could not therefore be said to be made by a definite act. Nor again, is there any sharp well-defined boundary set between light and darkness, so that we can say, "Here light begins, here darkness ends. " The very opposite is the case, the one blends imperceptibly into the other. This then cannot be the meaning of the words. But the next verse guides us to the real meaning. "And God called the light Day, and the darkness He called Night. " The division of light from darkness then is the alternation of night and day. When God divided the light from the darkness He made provision for that alternation. But we know that that alternation is the result of the earth's rotation upon its axis, so that the dividing the light from the darkness evidently implies the communication to the accumulated mass of the motion of rotation.

It does not clearly appear in the account of the first day, whether this alternation of day and night took effect immediately. Certainly the introduction of it here does not prove that it did so follow. For there was no way in which the fact of the earth's rotation could be directly communicated to those for whom the narrative was primarily intended. They were ignorant of the spherical form of the earth, and so could not have attached any idea whatever to a statement that it revolved about its axis.

The only way then in which Moses could speak of that rotation was in connexion with some phenomenon resulting from it. The only such phenomenon with which the Jews were acquainted was the alternation of day and night. There was therefore no way in which Moses could record the fact except with reference to this ultimate effect. It does not follow that that effect was immediate. Beside the rotation of the earth, another condition is required. The light must come from a single source, and so when the act is recorded by which that condition is effected, the division of light and darkness is again noticed. The sun and the moon are set in the firmament of heaven to divide the light from the darkness. But that division was potentially effected when the motion of rotation was given.

The third defect noticed in the Nebular Hypothesis was, that it did not account for this motion of rotation. This defect, then, like the two preceding ones, is supplied by the Mosaic Record, and the hypothesis thus supplemented becomes complete. It is capable of giving a satisfactory account of the phenomena to which it applies. But as it is only a theory, and only points out a way in which the universe might have been constructed, it does not in itself exclude the possibility that some other plan might in fact have been adopted, and we have now to examine into the reasons for supposing that it was the method which was actually employed. These divide themselves into two classes: —those which render it probable that similar processes are now in progress; and those which render it probable that the solar system has passed through such a process.

It has already been pointed out that the great nebulae in Orion and Argo seem to represent the condition of our system on the first appearance of light, and that changes are strongly suspected to be taking place in both; but we cannot expect to trace any single nebula through the stages of its development, since that development must occupy untold ages. All we can do is to inquire if there are other nebulas which seem to be in more advanced stages. It must at once be recognized, that if this be one of the processes now going on, it is not the only one. There are many nebulas "which have assumed forms for which the law of gravitation, as we know it, will not enable us to account—such as the Ring Nebula in Lyra, the Dumb-bell Nebula in Vulpecula, or the double Horseshoe in Scutum Sobieski. But some nebulas can be found which arrange themselves so as to illustrate the stages through which we may suppose our world to have passed. These are chiefly to be found among the planetary nebulse, which in a small telescope exhibit a faint circular disc, but in larger instruments frequently show considerable varieties of structure. Some of them present the appearance of a condensation of light in the centre, which gradually fades off; in others there is a bright ring surrounding the central spot, but separated from it by a darker space. The Nebula Andromeda 49647, [Footnote: The numbers are those given by Sir J. Hersohel.] as seen in Mr. Lassel's four-foot reflector appears as a luminous spot, surrounded by two luminous rings, which, in the more powerful instrument of Lord Bosse, combine into a spiral. Its spectrum is gaseous, with one line indicating some element unknown to us. In another nebula, Draco

4373, there is a double spectrum, the one gaseous, indicating the presence of hydrogen, nitrogen, and barium; the other, apparently from the nucleus, continuous, and so representing a solid or fluid mass, but so faint that the lines belonging to particular elements cannot be distinguished. [Footnote: Hugging, Philosophical Transactions, 1864.] Bridanus 846, and Andromeda 116, are probably similar nebulee occupying different positions with reference to us. They both give a continuous spectrum. The one in Bridanus is described as "an eleventh magnitude star, standing in the centre of a circular nebula, itself placed centrally on a larger and fainter circle of hazy light. " [Footnote: Lassell, quoted in Webb's "Celestial Objects, " p. 227.] The nebula in Andromeda assumes a lenticular form; that in Bridanus would probably present the same appearance if we saw it edge-ways. The former has probably increased in brilliancy in the course of centuries. Mr. Webb remarks of it, "It is so plain to the naked eye that it is strange the ancients scarcely mention it. " [Footnote: Webb's "Celestial Objects, " p. 180.] In these two nebulas we may perhaps see the mass ready to break up into separate worlds, the lenticular form being a natural result of extremely rapid rotation. Prom the fact that Andromeda 116 gives a continuous spectrum, Dr. Huggins inclines to the belief that it is an unresolved star cluster. But the reasons which led Sir W. Herschel to conclude that the nebula in Orion was gaseous, (a conclusion which, though for a time discredited by the supposed resolution of the nebula in Lord Kosse's telescope, was ultimately found to be correct), are equally applicable here. In general a certain proportion exists between the telescopic power requisite to render a star cluster visible as a nebulous spot, and that which will resolve it into stars; but this nebula, like that in Orion, though visible to the naked eye, cannot be resolved by the most powerful instruments yet made. And the nebula in Draco 4373, seems to present an intermediate stage between the purely gaseous nebula and this one. The faint continuous spectrum is probably the result of incipient central condensation. This nebula, if recent observations by Mr. Gill, of Aberdeen, are confirmed [Footnote: Popular Science Review, 1871, p. 426.], is much nearer to us than any of the fixed stars.

"We come now to the reasons derived from the Solar System itself, and of these there are several, some of them of considerable weight. The first is to be found in the uniform direction of almost all the motions of the system. They are from west to east. The sun rotates upon his axis, the planets revolve about the sun and rotate upon their axes, and the satellites, with one exception, revolve about their primaries, and, so far as is known, rotate upon their axes in the same direction, from west to east, and the motions take place very nearly in the same plane—the ecliptic. This seems to point to the conclusion that these motions have a common origin, as would be the case if all these bodies at one time existed as a single mass which revolved in the same direction. The one exception is to be found in the satellites of Uranus, whose motion is retrograde. But there are certain phenomena, which lead to the conclusion, that, on the outskirts of our system, there has at some time or other been an action of a disturbing force, of which, except from these results, we know nothing. "

[Footnote: Bode's "Law of Planetary Distances, " What holds good as far as Uranus, breaks down in the case of Neptune. Both Leverrier and Adams were to some extent misled by this law. The new planet should according to their calculations, based on this law, have been of greater magnitude and at a greater distance than Neptune.

The polar axis of Uranus, instead of being nearly perpendicular to the ecliptic, as in the case of all the other planets (except Venus), is nearly coincident with it. Venus occupies an intermediate position, the inclination of its equator to its orbit being 49 degrees 58'.]

There is also strong reason for believing that the sun is still a nebulous star, that the whole of the original nebula is not yet gathered up in the vast globe which at ordinary times is all that we can see. This aspect of the case, however, will come more fully under our notice when we come to the work of the fourth day. The figure of the earth, which is that naturally assumed by a plastic mass revolving about its axis, and the traces which it retains of a former state of intense heat, are both in accordance with this theory.

When these facts are duly weighed, there seems to be a reasonable probability that this process is the one which was actually employed in the formation of the solar system. The remarkable manner in which the theory adapts itself to the Mosaic account, and the fact that that account records special interferences of the Creator exactly at the points where the theory shows that such interferences would be necessary, give rise to a very strong presumption in its favour. We have in it also a clear illustration of the combination of general laws of nature with special interferences of Creative Power—the law of gravitation was called into action, and the work would proceed steadily under that law for a considerable period, till matters were ripe for a farther stage in the progress, and then the special interference would take place, in this instance the imparting the motion of rotation, and the work would again proceed under the natural law. All this while, however, the work would be one, and performed by one power, the only difference being in the direct or indirect action of that power.

The only point an reference to the first day which remains to be inquired into is the extent to which the work had proceeded at its close. As the commencement of the second day's work implies that at that time the earth had an independent existence, we may conclude that the first day's work comprehended the casting off of the several successive rings, and the condensation of those rings, or some of them, into the corresponding planets and satellites. These would probably still retain their intense heat, in virtue of which they would be luminous.

Many of the multiple stars may not improbably present to us much the same appearance as the solar system then presented. In many cases we have one large star, with one or more very minute attendants. Such a star is Orionis, a tolerably conspicuous star, which has two companions invisible to the naked eye, but visible with moderate telescopic power. (A telescope of 2.1 inches aperture, by Cooke, shows them well.) Five more companions are visible in a 4-inch telescope. In the large telescope at Harvard no less than 35 minute stars have been seen in apparent connexion with the brilliant star Vega. In all these cases it is true that the distances and periods of the companion stars are very much greater than in the case of the earth; but then our telescopes will only enable us to discern the more distant companions. Any small companion stars holding positions corresponding to those of the four interior planets, would be lost in the light of the primary star; and if, as is suspected, all the heavenly bodies are subject to some resistance, however small, from the medium in which they move, this resistance would in the course of ages diminish the mean distance, and with it the periodic time of the companion stars.

The latter part of the 5th verse has already been considered, and there is no need to recur to it at this point. At the close of the history we shall be in a better position to ascertain if any light has been thrown on that mysterious subject.

SECTION 5. THE SECOND DAY.

"And God said, Let there be a firmament in the midst of the waters, and let it divide the waters from the waters.

"And God made the firmament, and divided the waters which were under the firmament from the waters which were above the firmament, and it was so.

"And God called the firmament Heaven, and there was evening and there was morning, a second day"

The work of the second and third days evidently has its scene on the earth alone. At its commencement the earth appears to have become distinctly separated from the gradually condensing mass of the solar system, and to have assumed its spherical form. It had, in fact, acquired an independent existence; but it was still in a chaotic state. Its elements, which were hereafter to assume the three forms of solid, fluid, and gas, seem to have been still blended together. Of the three states, fluidity seems to have been that to which the mass most nearly approached. This seems to be indicated by the application of the term, waters, to the two parts into which it is now divided; for the Hebrew has no general word for "fluid, " so that the only method of expressing it was by the use of this word "water" in an extended signification; and all scientific investigations point to the same conclusion. The heat, as yet, must have been so intense that no rocks or metals with which we are acquainted could have remained in a solid form. The sorting out and first arrangement of the materials of the earth, with probably the farther development of a large portion of them by the introduction of a new element, seems to have been the work of the second day.

When we proceed to examine the narrative more closely, two important questions suggest themselves: —1. What special interference of Creative Power does it indicate? 2. What is the meaning of the division between the waters which were above the firmament and the waters which were under the firmament?

1. What special interference of Creative Power took place on the second day? Till within the last ten years, it would have been difficult to give a satisfactory answer to this question; for if all the elements were already in existence at the commencement of the second day, their arrangement would, as it seems, have been

brought about by the ordinary operation of natural laws which were already established. The cooling and condensation of a portion of the elements would have been effected by the radiation of their heat, and the portions thus condensed would, under the influence of gravitation, have arranged themselves in immediate proximity to the centre of gravity, forming a solid or fluid nucleus, round which those portions which still remained in a gaseous state would have formed an atmospheric envelope. But here again the spectroscope comes to our aid. In many of the nebulae which give in it the bright lines indicative of gas, hydrogen and nitrogen are the chief gases discovered. These must be in an incandescent state, or they would not be visible at all. But hydrogen cannot, in the present state of things, remain in this condition in contact with oxygen; it must instantly combine with it, that combination being attended with intense heat, and resulting in the production of water. The introduction of oxygen, then, must involve a very important crisis in the process of development; but that introduction must have preceded the formation of atmospheric air and water. Prior to the second day oxygen must either have been non-existent, or it must have existed in a form and under conditions very different from those under which it exists now. Free oxygen cannot be in existence in the sun or in any celestial object in which the spectroscope indicates the existence of incandescent hydrogen. The special act of the second day would appear to have consisted in the development of oxygen, or the calling it from a quiescent state into active operation.

But the effects of the new element thus called into operation would not be limited to the production of air and water. It is estimated that oxygen constitutes, by weight, nearly half of the solid crust of the earth. It forms a part of every rock and of every metallic ore. The second day, then, must have been a period of intense chemical action, resulting from the introduction of this powerful agent.

But (2) what is the meaning of the division of the waters which are above the firmament from the waters which were under the firmament? At present all the water contained in the atmosphere, in the shape of vapour and clouds, is so insignificant in comparison with that vast volume of water which not only fills the ocean, but also permeates the solid earth, that such a notice of it seems unaccountable. Mr. Goodwin, indeed, maintains that there was an ancient belief, not only that the firmament was a solid vault, but that on it there rested another ocean, at least as copious as that with which we are acquainted. [Footnote: Essays and Reviews, p. 220] In support of this assertion he brings forward the phrase, "The windows of heaven were opened" (Gen, VII. 11) and other similar expressions. But such phrases as this evidently belong to the same class as the fanciful names so often given to the clouds in the hymns of the Rig Veda. Both expressions evidently point to a time when figurative language, if no longer a necessity, was at all events a common and favourite form of speech, and was understood by all. Dr. Whewell [Footnote: Plurality of Worlds, chap. x. Section 5.] has put forward the curious notion that when the creation of the interior planets was completed, there remained a superfluity of water, which was gathered up into the four exterior planets. But the only fact in favour of such an hypothesis is the close correspondence between the apparent density of these planets and that of water. Now, as will be seen immediately, there is strong reason to believe that the true density of these planets is much greater than their apparent diameters would seem to indicate; so that the one solitary ground on which the suggestion rests vanishes when it is examined. Apart from this, however, the suggestion that there would be any superfluous material when the work of creation was finished, is a very strange one. Neither of these views, then, can be accepted as giving a satisfactory meaning to the text.

Astronomical investigations however, which have been carried on with great diligence during the last four winters, and which are still being continued with unremitting interest, have brought to light phenomena which seem to be in remarkable correspondence with the state of things spoken of in the text. It has already been noticed that the eight greater planets at present known to us are divided into two groups of four by the intervening belt of minor planets. These two groups have totally distinct characteristics. In density, magnitude, and length, of day the members of each group differ little from each other, while the two groups differ very widely. The moon is the only satellite as yet known in the inner group. The planets of the outer group are attended by at least seventeen satellites.

Of these outer planets Jupiter, from his great brilliancy, specially attracts observation, while from his comparative proximity to the earth we are enabled to examine him much more satisfactorily than we can Saturn, Uranus, or Neptune. Two facts with reference to him have long been well known, the one, that the polar compression in his case is much greater than it is in any of the interior planets, so that when seen through a telescope of very moderate power his disc is evidently elliptical, while the compression of the interior planets can only be detected by the most delicate micrometrical measurements—the other, that his apparent surface is always crossed by several alternating belts of light and shade, which though subject to constant changes of detail, always preserve the same general character. Until recently the generally received theory was that these belts consisted of clouds, raised by the heat of the sun, and arranged in zones under the influence of winds similar in character to, and produced by the same causes as, the trade-winds which blow over our own oceans. This view, however, has been shown by Mr. Proctor to be untenable. [Footnote: See a paper by Mr. Proctor in the Monthly Packet for October, 1870.]

About forty years ago, a very remarkable phenomenon was observed simultaneously, but independently, by three astronomers, Admiral Smyth, Mr. Maclean, and Mr. Pearson, who were watching a transit of Jupiter's second satellite from stations several miles apart. Admiral Smyth's account of it is as follows: —"On Thursday, the 26th of June, 1828, the moon being nearly full, and the evening extremely fine, I was watching the second satellite of Jupiter as it gradually approached to transit the disc of the planet. My instrument was an excellent refractor of 3 3/4 inches aperture, and five feet focal length, with a power of one hundred. The satellite appeared in contact at about half-past ten, and for some minutes remained on the edge of the limb, presenting an appearance not unlike that of the lunar mountains which come into view during the first quarter of the moon, until it finally disappeared on the body of the planet. At least twelve or thirteen minutes must have elapsed when, accidentally turning to Jupiter again, I perceived the same satellite outside the disc. It was in the same position as to being above a line with the lower belt, where it remained distinctly visible for at least four minutes, and then suddenly vanished. " A somewhat similar phenomenon, but of shorter duration, was witnessed by Messrs. Gorton and Wray, during an occultation of the same satellite, April 26, 1863. In this case the satellite reappeared after passing behind the apparent disc of the planet. So lately as 1868 this phenomenon was regarded as inexplicable. [Footnote: Webb's Celestial Objects, p. 141.]

In the winter of 1868-9 the attention of astronomers was called to the fact that rapid and extensive changes were taking place in the appearance of Jupiter's belts, and they have consequently been watched from that time with unremitting attention by astronomers furnished with telescopes of the best quality. The results of these observations are given in two very interesting papers, communicated to the Popular Science Review, by Mr. Webb. [Footnote: Popular Science Review for April, 1870, and July, 1871.] Very curious markings and variations in the depth of shade have been seen, accompanied by equally curious changes of colour. Mr. Browning compares these changes to those which are seen when a cloud of steam of varying depth and density is illuminated from behind by a strong light, as when we look through the steam escaping from the safety-valve of a locomotive at a gas-lamp immediately behind it. This appears to be the true explanation of the phenomenon. [Footnote: Popular Science Review, 1871, p. 307.] These belts are probably due to vast masses of steam, poured forth with great force from the body of the planet. As the atmosphere of Jupiter is probably of enormous depth, the rotatory velocity of its upper portions would be much greater than that of the surface of the planet, hence the steam would arrange itself in belts parallel to the equator of the planet. But this view leads us to wonderful conclusions with reference to the condition of the planet.

"Processes of the most amazing character are taking place beneath that cloudy envelope, which forms the visible surface of the planet as seen by the terrestrial observer. The real globe of the planet would seem to be intensely heated, perhaps molten, through the fierceness of the heat which pervades it. Masses of vapour streaming continually upward from the surface of this fiery globe would be gathered at once into zones because of their rapid change of distance from the centre. That which is wholly unintelligible when we regard the surface of Jupiter as swept like our earth by polar and equatorial winds, is readily interpreted when we recognize the existence of rapidly uprushing streams of vapour. " [Footnote: Mr. Proctor in Monthly Packet, October, 1870.]

Supposing then that the atmosphere of Jupiter is of very great depth, and thus laden with masses of watery vapour, the effect of a sudden current of heated, but comparatively dry, air or gas would be the immediate absorption of the whole or a large portion of the vapour, and the consequent transparency of the portion of the atmosphere affected by it. We see this result continually on a small scale in our own atmosphere, when a heavy cloud comes in contact with a warm air current, and rapidly melts away, Many of the rapid changes which have been witnessed in Jupiter's appearance are readily explained if this view is admitted. Supposing such a thing to have happened near the edge of the disc, the phenomenon recorded by Admiral Smyth is at once satisfactorily explained. When the satellite appeared to pass on to the disc, and to be lost in the light of the planet, it would for some time, proportional to the depth of Jupiter's atmosphere, have behind it a background of clouds only, it would not have entered upon the actual disc of the planet. If then these clouds were suddenly absorbed, the atmosphere behind the satellite would become transparent and invisible, the background would be gone, and the satellite would reappear. In the case of the occultation witnessed by Messrs. Gorton and Wray, the satellite would at first be hidden by cloud only, and would reappear if the cloud were removed. Such seems to be the true explanation of these hitherto mysterious phenomena. That they could not have resulted from any alteration in the motions of the planet or the satellite is evident. Such an alteration would have been instantly detected, since the places of both the planet and the satellites are computed years in advance, and any such change would at once have thrown out all these computations.

Assuming that this is the true solution of the mystery, we are enabled to form an approximate estimate of the extent of the atmosphere of Jupiter. The time between the first and second disappearances does not seem to have been accurately noted. Admiral Smyth's account makes it 16 or 17 minutes; but if we estimate it at 15 minutes only, and if we further assume that the second disappearance was upon the actual disc of Jupiter, and not upon a lower stratum of clouds, we shall be safe from any risk of exaggeration. The probability seems to be that the second disappearance was caused not by the disc, but by the formation of a fresh body of cloud, as it was not gradual, as in the first instance, but sudden. We shall then only have an estimate which cannot be greater, but may be much less, than the true value.

The mean distance of the second satellite from the centre of Jupiter is in round numbers 425,000 miles, and consequently the circumference of its orbit is 2,671,000 miles. The satellite travels through this orbit in about 86 hours, which gives a horary velocity of 31,400 miles, or 7850 miles in 15 minutes. This then is the least possible depth of the atmosphere of Jupiter. [Footnote: For the direction of the motion of the satellite would be at right angles to the line of sight.] The whole diameter of Jupiter, atmosphere and all, is 85,390 miles. Deduct from this 15,700 miles for the atmosphere, and we have for the diameter of the solid nucleus rather less than 70,000 miles. The height of the atmosphere is therefore not less than three-fourteenths of the radius of the planet, and may be much greater. The extent of the atmosphere, combined with the rapidity of rotation, accounts satisfactorily for the great apparent polar compression of the planet. Another inference is that the density of the planet must exceed the ordinary estimate in the proportion of two to one.

But next, the atmosphere of Jupiter is probably of very great density. Dr. Huggins states that he has observed in the spectrum of Jupiter "three or four strong lines, one of them coincident with a strong line in the earth's atmosphere. " [Footnote: Lecture at Manchester, November 16, 1870.] Strong lines mark increased density in the absorbent medium, and lines hitherto unobserved indicate new elements. It is therefore probable that the atmosphere of Jupiter is not only much more dense than that of the earth, but also contains some elements—which are absent from the latter. When with this fact we connect the very great extent of the atmosphere, it will be evident that the pressure at the surface of the planet will be enormous, and from this we can form an estimate of the intensity of the forces which must be at work in the interior of the planet, to project jets of vapour through such an atmosphere to so great a height.

The link which connects Jupiter with the earth, in the second stage of its existence, is the mention by Moses of the "waters which were above the firmament. " Viewed in the light of the present condition of the earth such a notice seems unaccountable. But if the earth at that time were in a condition similar to that in which Jupiter appears to be now, the water in the atmosphere or above the firmament would be a very important element in any description that might be given of it. It is in fact most probable that all the water (in the strict sense of the word) then in existence would be in a state of vapour, and that the waters which were under the firmament were the molten materials which afterwards formed rocks and ores, since, as has been already noticed, the word is the only one which could be employed to describe fluids in general.

We may now try to form some idea of the probable state of the earth at this period. Its centre would be occupied by a fused mass, in which were blended all the more intractable solid constituents of the present world. This would be surrounded by an atmosphere of very great height and density, containing not only all the present constituents of air, but also all, or nearly all, the water, and all the more volatile of the metals and other elements. Carbonic acid, to a very large extent, would probably be present, and a very considerable proportion of the oxygen which now exists in combination with various bases, and forms by weight so large a proportion of the solid crust of the world.

Owing to the intense heat, chemical combinations would readily be formed between the ingredients of the fused mass and the other elements which existed in the form of vapour, and thus the earliest of the vast variety of existing minerals would be elaborated. The volumes of steam which floated in the upper regions of the atmosphere would rapidly part with their heat by radiation into space, and would descend towards the surface of the earth in the form of rain. At first probably, and for a long time, they would not reach the surface, but as they approached it would be again converted into vapour, and re-ascend to pass again and again through the same process. But by this means the intense heat of the nucleus would be gradually conveyed away, till the cooling reached a point at which some of the superficial materials would assume a solid form. It is by no means certain what is the true primary rock— for a long time it was almost universally assumed to be granite, since granite is uniformly found underlying the oldest sedimentary rocks that are known. But as these rocks have been forced from their original position and tilted up, the underlying stratum may probably be of later date than the upper ones, since it was the elevating agent. So that we can have no certain knowledge on this point, since the earliest sedimentary strata, wherever they retain their original position, must be at a depth far below the reach of man. If, however, Sir C. Kyell's view of the conditions requisite for the formation of granite are correct, these conditions [Footnote: Student's Geology, chap. xxxi.]—heat, moisture, and enormous pressure—would all be present at the surface of the nucleus. Some kind of solid floor must have been formed before the next stage could be reached, at which it would be possible for water to exist in a fluid state. This, however, would be possible at a much higher temperature than at present, owing to the enormous atmospheric pressure. It is possible now, by artificial means, to raise water, nearly if not quite, to a red heat, without the formation of steam, and the pressure of the atmosphere in the case supposed would, in all probability, be much greater than any which we can now apply under the conditions necessary for heating the water.

It is probable that at this point the close of the second day must be placed: but the indications of the narrative do not enable us to fix it with any degree of certainty. As, however, from this point a new series of processes would commence, and those processes are in intimate connexion with the first of the two developments ascribed to the third day, the period when water could first maintain a fluid form on the earth's surface, seems to present the most probable line of demarcation.

SECTION 6. THE THIRD DAY.

"And God said, Let the waters under the Heaven be gathered together in one place, and let the dry land appear; and it was so.

"And God called the dry land Earth, and the gathering together of the waters called He Seas, and God saw that it was good.

"And God said, Let the earth sprout sprouts, the herb seeding seed, and the fruit-tree yielding fruit after his kind, whose seed is in it, [Footnote: "It" seems preferable to "itself" here. The same Hebrew word stands for both, but if the "fruit-tree" be taken as the antecedent, which it must be if we translate "itself, " there seems no meaning in the statement. If we read "it, " the pronoun will refer to the fruit—"the tree whose seed is in its fruit"—which gives an intelligible sense.] upon the earth, and it was so.

"And the earth caused to go forth sprouts, the herb seeding seed, and the fruit-tree yielding fruit whose seed is in it, after his kind, and God saw that it was good. And there was evening, and there was morning, a third day. "

The record of the third day is a very important one, because it is the first point at which the Mosaic Record comes in contact with that other record which is written in the rocks. Up to this time we have only been able to compare the statements of Moses with conjectural views of the earliest condition of the earth, which, though they may be highly probable, are at best only conjectures. But from this point we have to deal with a number of ascertained facts—certain landmarks stand out which enable us to fix the correspondent parts of the two narratives, and guide us to the identification and interpretation of their minor details.

The first of these landmarks is the appearance of the dry land, or, in geological language, the commencement of the process of upheaval. At the close of the second day the earth was, in all probability, as we have seen, a globe internally molten, but having a solid crust which was uniformly covered with a layer of water, and surrounded by an atmosphere which, though it had parted with some of its ingredients, was still very much more complex, more dense, and more extensive than it is at present. The newly condensed waters would rest on the surface of the primeval rock, whatever that rock might be. The internal heat conducted through it would keep the waters in a state of intense ebullition, and at the same time their surface would be agitated by violent atmospheric currents as the heated air ascended, and was replaced by cooler air from the outer regions of the atmosphere. Under these circumstances the water would dissolve or wear down portions of the newly-formed rock on which it rested. At the same time the steam, which would be continually rising from the boiling ocean, would descend from the upper regions of the atmosphere in the form of rain, and bring with it in solution considerable quantities of those elements which still existed in the form of vapour, just as rain now brings down ammonia and carbonic acid which it has absorbed in its passage through the atmosphere. New combinations would thus be formed between the materials dissolved or abraded by the ocean and those brought down by the rain. When these combinations had reached a certain amount they would be deposited in the form of mud upon the bed of the ocean, and thus the earliest sedimentary rocks would be formed. As the temperature gradually decreased, the character of these combinations would probably be changed, and at the same time the atmosphere would be diminished in volume and density, and become more pure by the absorption of a large portion of its original constituents, which would have been incorporated into various minerals.

The earliest sedimentary rock with which we are acquainted at present is what is known as the Laurentian formation. [Footnote: The whole of the geological details in this section are taken from Sir C. Lyell's Geology for Students.] It occupies an area of 200,000 square miles north of the St. Lawrence; and is also traced into the United States and the western highlands of Scotland and some of the adjacent isles. It is divided into two sections—the Upper and Lower Laurentian. It is not certain that it is really the oldest rock; for as every sedimentary rock is formed of the debris of preceding rocks, it is very possible that all the exposed portions of some older rocks may have been decomposed and worn away; but it is the oldest yet known. The thickness of the lower portion is estimated at 20,000 feet, or nearly four miles, while the Upper Laurentian beds are 10,000 feet thick. At this point we meet with the first traces of that process of upheaval and subsidence which has ever since been going on in the earth. The Lower Laurentian rocks had been displaced from their original horizontal position before the Upper Laurentian were deposited upon them.

This process of upheaval of some parts of the earth, accompanied with subsidence in other parts, is one which cannot be accounted for by any natural laws with which we are acquainted. It is in all probability the result of a series of changes which are taking place in the interior of the earth, but of which we know nothing at all. It is in the commencement of this series of changes that we trace that direct interference of the Creator—which is indicated by the command, "Let the waters under the firmament be gathered together into one place, and let the dry land appear. " We have not, however, any means of ascertaining how long a period elapsed before the process of upheaval reached the point at which the land would rise above the surface of the ocean.

The Lower Laurentian rocks are remarkable in another way. There is little doubt that traces of life, the earliest yet known, occur in them. They include a bed of limestone varying in thickness from 700 to 1500 feet. In all probability limestone, wherever it occurs, is an animal product, though in many cases all traces of its organization have been lost by exposure to heat. This particular bed appears to have been formed by a very lowly creature, which in organization was akin to the foraminifera, of which large quantities are now known to exist at the bottom of the Atlantic. It differed from them, however, in one respect—the individuals were connected together, as is the case now with many varieties of the coral animal. No notice of this first appearance of life is found in the Mosaic Record, nor, for reasons already given, was it possible that any mention of it should be made.

The rocks which come next to the Laurentian in the order of time are those known as the Cambrian. They are so called because they constitute a large portion of the mountains of North Wales, and it was there that their characteristics were first carefully studied by Professor Sedgwick. In one of the strata of this formation—the Harlech Grit—what are known as "ripple-marks" are found, proving that parts of these rocks at the time of their deposition formed a sea-beach, and that consequently at this time, at the latest, the dry land had emerged from the ocean. In these rocks there are also decided traces of Volcanic Action, which seem to indicate the existence of a Volcano similar to the recent "Graham's Island. " At this point a considerable advance in animal life is found. The fossils comprise several corals, varieties of mollusca, and a class of crustaceans peculiar to the very early rocks—the trilobites.

On the Cambrian rocks rest the formations known as Silurian, from the fact that they were first thoroughly examined in South Wales (Siluria) by Sir E. Murchison. In these rocks many fresh varieties of invertebrate fossils are found, and the vertebrata make their first appearance, numerous remains of fishes having been discovered. The earliest specimen was found in the Lower Ludlow beds at Leintwardine, while the Upper Ludlow formation contains an extensive bed composed almost entirely of fish-bones. Immediately above this bed are found what seem to be traces of land-plants, in the shape of the spores of a cryptogamous plant.

The Silurian rocks are succeeded by rocks which present two distinct characters, but are probably contemporaneous, the Devonian and the old Red Sandstone. The former seem to have been deposited in the bed of the sea, while the latter is a fresh-water formation. In these decided remains of land plants are found, of which about 200 species have at present been discovered. The old Red Sandstone is also peculiarly rich in fossil fish. The first signs of coal appear in this series of rocks, but on a very small scale.

We now come to what are known as the Carboniferous rocks, of which the lower series is known as the mountain limestone, and above it come the "coal measures, " containing numerous beds of coal, sometimes of great thickness. These beds have resulted entirely from the decomposition, under peculiar circumstances, of an enormous development of terrestrial vegetation. They seem to have originated in vast swamps, subject to occasional flooding, and to alternate movements of upheaval and subsidence. On these swamps there must have existed for ages a vegetation of whose luxuriance the richest tropical jungles of the present time can give us no idea. They tell the tale of a time when the temperature of the earth, was uniformly high (since coal fields are found in high northern latitudes), when the atmosphere was charged with moisture, and probably contained a large proportion of carbonic acid. In the coal measures we come upon the first traces of land animals. Several remains of reptiles have been found, as well as footprints left on the soft mud or sand of a riverbank or sea- beach. There seems to be no doubt that they were left by lung- breathing animals.

The carboniferous strata form the second of our landmarks. They seem to point to the fulfilment of the command that the earth, should bring forth vegetation. There is, however, one point which requires some notice. The Mosaic account, as we read it in our English Bibles, seems to be limited to phanerogamous plants— grass, the herb yielding seed, and the fruit-tree yielding fruit. Now, it is a well-known fact that the great mass of the vegetation, the remains of which constitute coal, consisted of cryptogamic plants, which do not produce seed, properly so called, but only spores; the distinction being that the spore contains the germ and nothing more, while in the seed the germ is provided with a store of nutriment to assist in the earlier stages of the development of the plant. What appears to be a farther discrepancy, the absence of any traces of the grasses, leads in reality to the solution of the difficulty.

The word which is translated "grass" [Hebrew script] means in reality, any fresh sprout. Now it is remarkable that Moses specifies three kinds of vegetation, with regard to two of which it is noted that they produce seed, while nothing is said of the seed of the remaining class. Grass too, is really a herb bearing seed, and, as such would be included in the second class, and there would have been no occasion, to mention it separately. It would appear then that the first class consisted of seedless plants, i. e. of the cryptogamia. This conclusion is strengthened when we turn to verses 29 and 30. If the word [Hebrew script] were correctly translated "grass, " we should certainly expect to find it in those verses, since the grasses contribute more to the food of both man and beast, than all the other herbaceous plants put together. This omission then, is an indication that the word, as used in this chapter, denotes a class of plants which are not commonly employed for food, and this condition also is fulfilled in the cryptogamia.

There are then four special points in this period, of which two seem to correspond with the Mosaic record, while the other two are unnoticed in it. The two points of correspondence are the upheaval of the dry land, and the prevalence of a very abundant and luxuriant Flora. As in the case of the fifth and sixth days, the words used with reference to land plants seem to denote a period of remarkable development, rather than the first appearance. The two points unnoticed are the beginnings of animal and vegetable life. In the case of animal life the omission has already been accounted for. The beginning of vegetable life was probably contemporaneous with that of animal life, for each is necessary to the other, since the food of the animal must be prepared by the vegetable, and after being used by the former returns to a state in which it is fitted for the nourishment of the latter. As animal life commenced in the ocean, so in all probability did vegetable life, though no certain traces of it are found in the earliest rocks; but this is easily accounted for by the very perishable character of the simpler forms of algae. Like the earliest animals, the first algae were probably microscopic plants, and the omission of any mention of them was therefore inevitable.

One characteristic of cryptogamic vegetation is important for its bearing on the work of the fourth day. Almost all the phanerogamic plants are dependent for their development upon the direct light and heat of the sun. Deprived of these they either perish entirely, or make an unhealthy growth, and produce little or no fruit. But the cryptogamia, in general, thrive best when they are protected from the direct rays of the sun. They nourish in a diffused light, and with abundant atmospheric moisture. And so we find them at this time doing what seems a very important work in the progress of the world. By taking up and decomposing the excess of carbonic acid which at this time probably existed in the atmosphere, they at once purified that atmosphere, and rendered it fit for the respiration of more highly organized creatures, and laid up in the earth an invaluable store of fuel for the future use of man. The other orders of vegetation seem to have existed in very small proportions at this time, and only in their lower forms. As the conditions of the earth changed, the cryptogamia seemed to have dwindled away, while higher forms of vegetation asserted their supremacy. It is not, however, improbable that a special development at a much later period is indicated by the mention in the second chapter of the formation of the garden of Eden.

SECTION 7. THE FOURTH DAY.

"And God said, Let there be luminaries in the firmament of heaven to divide between the day and the night, and let them be for signs and for seasons, and for days and for years.

"And let them be for luminaries in the firmament of heaven to give light upon the earth; and it was so.

"And God made the two luminaries, the great ones; the luminary, the great one, to rule over the day, and the luminary, the small one, to rule over the night, and also the stars.

"And God gave them in the firmament of heaven to give light upon the earth.

"And to rule over the day and over the night, and to divide between the light and between the darkness; and God saw that it was good.

"And there was evening, and there was morning, a fourth day. "

This day's work differs from that of the preceding and succeeding days, in the fact that its sphere was without the earth, which was only indirectly influenced by it, and consequently the geological records give us no direct information upon the subject, though in two points they tally with the Mosaical account. In the first place, the deposits of coal, which preceded this period, indicate a time when a nearly uniform temperature, and that a high one, prevailed throughout the world. The coal beds are found not only in tropical regions, but in very high latitudes. Not only is the vegetation of which these coalfields are the result, analogous to that which is now found in warm climates only—(this might be the case, and yet we should not be justified in drawing the inference that the actual species of plants were tropical, for it often happens that different species of the same genus, having considerable external resemblance, are very different in their habits, some requiring tropical heat, while others flourish only in temperate climates)—but the marked feature is the astonishing luxuriance of this vegetation, which could only have been developed under the most favourable circumstances of warmth and moisture. Now the heat which any particular portion of the earth's surface receives from the sun depends entirely upon the latitude. hence it is impossible that a uniform high temperature could exist in a world which derived its heat wholly or chiefly from that source. Whether the high temperature which prevailed on the earth during the deposition of the coal measures was derived from internal heat it is impossible to say; it is evident that the temperature of the earth's surface has been in past times, and perhaps is now, modified by causes which no scientific research has been enabled to detect [Footnote: Since the sun's secular motion has been known, astronomers have suggested that the solar system has been carried through portions of space having variable temperatures. Geologists, however, do not seem inclined to accept this as a sufficient reason for the phenomena observed.]. But we may safely conclude that during the third day the earth did not derive its heat from the sun. The second point, the barrenness of the geological records of this period, will be noticed hereafter.

The record of the fourth day's work admits of two interpretations, it may describe things merely as they appeared, or as they actually occurred.

1. It is possible that the events of the fourth day may be described phenomenally—that up to this period the state of things on the earth had been to a great extent similar to that which we have reason to believe is still existing in the planet Jupiter- that the atmosphere was so charged with vapour that no direct rays from the heavenly bodies could penetrate it; but that at this time, owing to the declining heat, a great part of the aqueous constituents of this vapour had been precipitated in the form of rain, while other vapours had entered into chemical combinations with other elements to form the various minerals of the earth's surface, and the atmosphere had become first translucent, and then transparent. While this process was going on, no direct light from the sun, supposing it to be already in existence, could penetrate the veil. Diffused light only could reach the earth's surface, but when the atmosphere became clear the sun, moon, and stars would become visible.

Against this view several objections may be brought. In the first place, as has been already noticed, we cannot treat the account of the Creation as derived from ordinary human sources. Either it is a revelation from the Creator or it is nothing. Now we can readily admit that a man, speaking of an event which lie had witnessed, but did not understand, would describe it as it appeared to him, but we cannot admit this supposition when the work is described by the Great Artificer Himself. In the next place, the temperature of the earth's surface must in this case have been affected by the sun, and must therefore have been more or less dependent upon latitude— and in the third place the distinction between day and night must have come into operation, whereas the narrative implies that it was yet incomplete.

2. The other possible interpretation is, that at this period the concentration of light and heat in the sun was so far completed that he became the luminary of the system, which had hitherto derived its light and heat from other sources. Probably, for a long time, the internal heat of the planets may have been so great that they were a light to themselves. This state of things, however, must have come to an end before animal or vegetable life could have existed on their surface, but other ways exist, and are in operation in other parts of the universe, by which light and heat might have been supplied independently of the sun. That light which is now gathered up in the sun might for a long time have existed as a nebulous ring, similar to the well-known Ring Nebula in Lyra. Any planets existing within such a ring would probably derive from it sufficient light and heat. Or the nebulous matter, in a luminous state, while slowly advancing to concentration, might as yet have been so diffused as to fill a space in which the earth's orbit was included. In either case the earth would have received a uniform diffused light, without any alternations of night and day. It is of course impossible that we should be able to say whether there are any worlds in which such a state of things prevails at present. Up to this time, with one possible exception, [Footnote: "Sirius is accompanied by a 10 mag. star, whose existence was suspected (like that of Neptune), long before its discovery by Alvan Clark in 1861, from the irregular movements of its primary. But though it appears so small, its disturbing effects can only be accounted for on the supposition that its mass is at least half that of Sirius, in which case its light must be very faint, possibly wholly reflected. " (Webb's Celestial Objects, p. 202.)] the only worlds which the telescope has revealed to us, beyond the limits of our own system, are self-luminous. No reflected light is strong enough to make its existence perceptible at such enormous distances in the most powerful telescope which has yet been constructed.

There are some facts connected with our own system which make it appear not improbable that up to the time of which we are speaking the light which is now gathered up in the sun was diffused over a space in which at all events the earth's orbit was included. It is now a recognized fact that all the light of the system is not as yet wholly concentrated in the sun, as we generally recognize it, but that to some extent the sun is still a nebulous star. Under ordinary circumstances we see only that circular disc, which we usually recognize as the sun. Its surpassing brightness overpowers every thing else, whether we view it with the unaided eye or through the telescope. But when the actual disc is hidden from us by the moon in a total eclipse, other regions of light surrounding the disc, make their appearance, and in them the most wonderful processes are continually going on. The simultaneous discoveries of Messrs. Lockyer and Janssen, in 1868, have enabled some of these processes to be continuously watched when the sun is not eclipsed, but others can as yet only be seen during the few minutes (never amounting to seven) which a total eclipse lasts, so that as yet we know very little of them.

Immediately surrounding the disc of the sun, which is visible to the naked eye, is a brilliant ring of light, known now as the chromosphere or sierra. This is the region which till 1868 could be seen only during total eclipses, but can now be watched at all times by means of the spectroscope. In it symptoms of intense action are from time to time witnessed. For many years past, whenever a total eclipse occurred, there were observed on the edge of this ring certain red prominences. The spectroscope has revealed their nature. They consist chiefly of enormous volumes of hydrogen, ejected from the surface of the sun with a velocity almost inconceivable, and at the same time revolving about their axis after the fashion of a cyclone. [Footnote: Popular Science Review, January, 1872, p. 150; Look. Byer's Lecture on the Sun, at Manchester, 1871.] A very remarkable instance of this was observed in America in September 1871, by Professor Young. A mass of incandescent hydrogen was propelled to a height of 200,000 miles above the visible disc; of these the last 100,000 miles were passed through in 10 minutes. Such events, though not commonly on so vast a scale, are continually occurring on the surface of the sun, and they seem to be in close connexion with the magnetic phenomena occurring on the earth.

Beyond the chromosphere lies the corona. The spectroscope has not yet rendered this visible at all times, and consequently we are dependent upon the information to be obtained during the few minutes of total eclipses, when alone it is visible. Consequently during recent solar eclipses this has been the point to which the attention of astronomers has been especially devoted. The eclipse of December, 1870, decided one point, that the corona was a truly solar phenomenon, and not, as some astronomers imagined, an optical phenomenon, produced by our own atmosphere. The corona presents the appearance of nebulous light, fading as it becomes more remote from the sun, of very irregular outline, at some points not extending more than 15', at others as much as 60' or 70' from the sun's disc, or, in other words, reaching to distances from the sun's surface varying from 400,000 to 1,800,000 miles. More important information has been obtained from the eclipse of December 12,1871. It is now ascertained that the corona comprises not only gaseous elements, especially hydrogen, but also solid or fluid particles, capable of giving a continuous though very faint spectrum with dark lines, indicating the existence of matter capable of reflecting light. The character of the coronal spectrum very much resembles that of the Nebula in Draco, No. 4373. The ascertained extent of the corona exceeds a million of miles above the surface of the sun, and it seems probable that the Zodiacal light is only a fainter extension of it. [Footnote: Popular Science Review, April, 1872, pp. 136-146.]

On a clear evening in the early spring months, as soon as twilight is completely ended, a conical streak of light may be sometimes seen, arising' from the western horizon, and extending through an arc of 60 or 70 degrees, nearly in the direction of the Ecliptic, and finally terminating in a point. This is the Zodiacal light. In tropical climates it is seen much more frequently, [Footnote: Humboldt, Kosmos, vol. i. p. 126 (Bohu's edition).] and is much more brilliant than in England. This then is probably an envelope of still fainter light than the corona. It must extend beyond the orbit of Venus, as the maximum elongation of Venus is 47 degrees, while the Zodiacal light has been traced for 70 degrees, and probably farther. It is very possible that the earth is occasionally involved in it, and that from it we derive that diffused light which, though faint, is very serviceable to us on a starless evening, and of which no other account has as yet been given. The light we receive in this way is often as powerful as that which we should receive from the stars if they were not hidden by clouds.

These phenomena seem to point to the conclusion that the condensation of light in the sun has been a very gradual process, which is even yet incomplete. If we suppose that at the time of the formation of the coal measures it was not far advanced, but that a diffused light extended beyond the orbit of the earth, similar in some respects to the present Zodiacal light, but equal in intensity to the light which we now see in the corona, the phenomena of the third day will be satisfactorily accounted for. There is, however, still an enormous amount of mystery connected with the sun. It is the centre from which an inconceivable amount of force in the shape of light, heat, actinism, and probably other manifestations, is hourly poured forth. If the whole of that force were divided into two thousand million parts, the portion received by the earth would be represented by one of those parts, and the whole amount received by all the planets would fall short of twelve of them. All the rest is radiated away into space, and so far as we know at present lost to the system. The question then arises, "How is this enormous expenditure supplied? " Various sources of heat have been suggested, but none of them seem satisfactory. One conceivable source there is, but that lies out of the domain of science. Then again, metals, which only our most powerful furnaces will even melt, exist in the sun's atmosphere in the state of vapour. What must be the intensity of the heat which underlies that metallic atmosphere? and what can be the solid or fluid substances which, from the continuity of the spectrum, we know must exist there?

We turn now to the Mosaic Record to see what light it throws upon and receives from this investigation. The first thing to be noticed is that the word used by Moses for the sun and moon is not the same as that employed to denote light. It properly signifies a light-holder, such as a candlestick, and harmonizes with the view that the sun in his original state was not luminous, but was made a luminary by the condensation of light previously existent under other conditions. In the next place, though the apparent dimensions of the sun and moon are the same, Moses correctly describes the one as "the great light, " the other as "the little light, " thus indicating a knowledge to which the astronomers of his day had probably not attained.

The relation between the accounts of the first and fourth day's work becomes clear if we assume that the sun was not made a luminary till the fourth day. The division of night and day depends upon two things, the rotation of the earth upon its axis, and the concentration of light in the sun. Hence when the rotation of the earth commenced that division was potentially provided for, but the provision would not take effect until the second condition was fulfilled by the concentration of light in the sun. The indications given by the coal measures point, as we have seen, to the same conclusion.

The only remaining question is "What was going on in the earth at the same time? " Our materials for answering this question are but scanty. So great an alteration in the sources of light and heat must have involved great physical changes on the earth's surface, and there is reason to believe that great mechanical forces were at work producing vast changes in the relations of land and water. "It has long been the opinion of the most eminent geologists that the coalfields of Lancashire and Yorkshire were once united, the upper coal measures and the overlying Millstone Grit and Toredale Bocks having been subsequently removed by denudation; but what is remarkable is the ancient date now assigned to this denudation, for it seems that a thickness of no less than 10,000 feet of the coal measures had been carried away before the deposition of even the lower Permian Rocks, which were thrown down upon the already disturbed truncated edges of the coal strata. " [Footnote: Lyell, Geology for Students, p. 377.] And this is but a single instance.

During the interval between the deposition of the coal measures, which seem to belong to the third, and the Saurian remains which mark the fifth day, we have the Permian and Triassic Rocks, of which the Magnesian. Limestone and the new Red Sandstone are the most important representatives in England. Till a very recent period it was thought that these rocks belonged to a period remarkably destitute of animal life, very few fossils having been found in them. Recently, however, some very rich deposits have been found in the Tyrol, belonging to this period, but they are only local.

Of the Permian formation Sir C. Lyell says, "Not one of the species (of fossils) is common to rocks newer than the Palaeozoic. " [Footnote: Geology for Students, p. 369.] This was not then a time for the origination of new forms of life. In the Trias, however, the new development of life, which was to attain its full dimensions on the fifth day, begins to open upon us. The earliest Saurian fossils are found, and the rocks still present us with impressions of the feet of reptiles and birds, which walked over the soft seashore, and left footprints, which were first dried and hardened by the sun and wind, and then filled up with fresh sand by the returning tide, but never entirely coalesced with the new material.

At the close of this period the first traces of mammalian life occur, in the shape of teeth, which are supposed to have belonged to some small Marsupial quadrupeds, and in America the whole lower jaws of three such animals have been discovered; but no other remains have as yet been traced.

The Trias then seems to mark the boundary between the fourth and fifth days. The fourth day seems to have been on the earth a period of great change, not only in physical conditions, but also in the forms of life. In the latter point of view, however, it seems to have been marked by the passing-away of old forms much more than by the origination of new ones, and hence the barrenness of the Geological

Records is in exact accordance with the silence of the Mosaic Record as to any new developments.

SECTION 8. THE FIFTH DAY

"And God said. Let the waters swarm swarms, the soul of life, and let fowl fly above the earth in the face of the firmament of heaven.

"And God created the monsters, the great ones, and every soul of life that creepeth, with which the waters swarmed, after their kind, and every winged fowl after his kind; and God saw that it was good.

"And God blessed them, saying, Be fruitful and multiply, and fill the waters in the sea, and let fowl multiply on the earth.

"And there was evening, and there was morning, a fifth day. "

The fifth and sixth days of Creation are those to which the theory of development chiefly refers. It will, therefore, be better to defer the consideration of its bearing on the narrative till the relation of that narrative to Geological facts has been considered, since it can only be thoroughly weighed when taken in connexion with the facts which belong to the two days.

The beginning of the fifth day may be assigned to a point near where the Trias is succeeded by the Lias. As the Trias is drawing to its close, the class of reptiles, whose first known appearance belongs to the carboniferous epoch of the third day, begins to show signs of advance. The first true Saurians are found in the Trias: the great development takes place in the Lias and Oolite, while in the chalk large quantities of kindred remains are found, which, however, are not identical with the species found in the earlier groups. Of these some were probably almost entirely aquatic, as their limbs take the form of paddles; others were purely terrestrial, a large proportion were amphibious, and some, as the pterodactylus, bore the same relation to the rest of their class as the bats bear to the other mammalia, being furnished with membranous wings, supported upon a special development of the anterior limbs. One important characteristic of the race at this time was the great size of many of its members: thirty feet is by no means an uncommon length. This marks the fitness of the name given to the class by Moses.

Very few actual remains of birds have been found; but this is not surprising, since birds would rarely be exposed to the conditions which were essential to the fossilization of their remains. The earliest known fossil bird is the Archaeopteryx, the remains of which were found in 1862 in the Solenhofen Slates, which belong to the Oolite formation. Though the actual remains of birds are very few, traces of their footprints have been found in many places, from the New Red Sandstone upwards, and these traces prove not only that they were very numerous, but also that they attained to a gigantic size, as their feet were sometimes from twelve to fifteen inches in length, and their stride extended from six to eight feet. During this period, then, these two classes must have been the dominant races of the earth. As the precursors of these classes made their appearance at a much earlier period, so the epoch of birds and reptiles witnessed the beginning and gradual advance of the class which was to succeed them in the foremost place—the mammalia. Generally, however, the mammalian remains of this period belong to what are considered the lower classes—the monotremata and marsupialia. The close of this period must have been a time of great disturbance in the Northern Hemisphere, since the chalk which runs through a great part of Northern Europe, and frequently attains a thickness of 1000 feet, must have been deposited at the bottom of a deep sea, and subsequently elevated.

SECTION 9. THE SIXTH DAY.

1. The Mammalia.

"And God said, Let the earth cause to go forth the soul of life, cattle, and creeping thing, and the beast of the earth (wild animals) after his kind; and it was so.

"And God made the beast of the earth after his kind, and cattle after their kind, and every creeping thing of the ground after his kind; and God saw that it was good. "

In these two verses there are one or two points which call for notice. In the first place, the creatures mentioned are divided into three classes, of which two, cattle and the beast of the earth, are tolerably clear in their general significance, though their extent is not determined. The third is denoted by a word which had already been employed to describe the work of the fifth day, and is translated in our version "creeping thing. " The probability seems to be that it has reference to such classes of animals as the smaller rodentia, and the mustelidas, whose motions may be appropriately described by the word "creeping. " That it denotes four-footed creatures has already been pointed out. The next point is, that in each case the singular is used; in the case of the domestic animals this fact is lost to the English reader by the use of the collective noun "cattle. " Of course it is a common usage, to denote a class of animals by a singular noun used generically, but the statements of the passage would also be justified if one pair only of each of the three types specified were called into existence at first. It is also to be noticed that while the word [Hebrew script], the earth is used to define the wild beast; another word, [Hebrew script] the ground, is applied to the "creeping thing. " There is probably a reason for this, though it may not at present be apparent.

When we turn to the Geological record, we find that the period of the chalk was followed by the deposition of the tertiary strata. During the upheaval of the chalk these strata seem to have been gradually laid down in its hollows, and around its edges. They extend from the London clay upward to the crag formations which appear on the Eastern coast of England at intervals from Bridlington to Suffolk. In these strata we see signs of an approach to the existing state of things. As we ascend through them, a gradually increasing number of the fossil shells are found to be specifically identical with those which at present inhabit the ocean.

Another characteristic of this period is the abundance of fossil remains of mammalia; but in this case, although the remains are evidently, in many cases, those of creatures nearly allied to those now existing, they are not identical, very great modifications both of bulk and of minor structural details having taken place. One very important point of difference is the vastly superior bulk of these ancient animals: a good illustration of which may be seen in the skeletons of the mammoth and of the modern elephant, which are placed near each other in the British Museum. Many of these animals appear not to have become extinct till long after the appearance of man.

The first appearance of mammalia, as has been already noticed, must have been long before this, as the earliest fossils yet found are at the lower limit of the Lias. They belong, however, to the genus Marsupialia, of which, as far as we know, no representatives were in existence in any part of the world known to Moses, so that even on the supposition that he intended to give an account of the first appearance of the classes of animals which he mentions, the omission of these would have been inevitable. His words, however, appear to point to a time when the mammalia occupied the leading place, just as the reptiles had occupied the leading place at a previous epoch. And his words are fully borne out by the records of the rocks.

At the close of the tertiary period great changes once more took place in the Northern hemisphere. There was a great and extensive subsidence, in consequence of which a large portion of Northern and Middle Europe must have been under water, the mountain summits only appearing as detached islands. At the same time, from causes utterly unknown to us, there was a great depression of temperature, the result of which was, that all, or nearly all the land, in those regions which were not submerged, was covered with glaciers, much as Greenland is now, and from these glaciers vast icebergs must from time to time have been detached by the sea and floated off, carrying with them fragments of rock, some freshly broken, some rounded by long attrition, which were deposited on the then submerged lands as the ice melted, and are now found as boulders, sometimes lying on the surface, at others dispersed through beds of clay and sand formed under water from the debris worn down by the glaciers. A subsequent movement of elevation ushered in the state of things which exists on the earth at the present time.

2. Man.

"And God said, Let Us make man (Adam) in Our image after Our likeness; and he shall have dominion over the fish of the sea, and over the fowl of the heaven, and over the cattle, and over all the earth, and over every creeping thing that creepeth upon the earth.

"And God created man (the Adam) in His image, in the image of God created He him; male and female He created them.

"And God blessed them, and God said to them, Be fruitful and multiply, and fill the earth and subdue it; and rule over the fish of the sea, and the fowl of the heaven, and over every animal that creepeth upon the earth.

"And God said, Behold, I have given you every herb seeding seed, which is upon the face of all the earth, and every tree which has in it the fruit of a tree seeding seed; to you it shall be for food.

"And to every animal of the earth, and to every fowl of the heaven, and to every thing that creepeth upon the earth, in which is the soul of life, every green herb is for meat; and it was so.

"And God saw every thing—which He had made, and behold it was good exceedingly.

"And there was evening and there was morning, the sixth day. "

The terms in which the Creation of man is spoken of are such as to challenge particular attention and to induce us to expect something very different from what occurred on any previous occasion. In the first place, more agents than one are introduced by the use of the plural form of the verb, and thus at the very commencement of man's career there is an intimation of that mysterious fact of the Trinity in Unity which was to have so important an influence upon his future destiny. Then we are told that man was to be formed in the Image of God, a statement which probably is of very wide import. It has been variously interpreted as having reference to the spiritual, moral, and intellectual nature of man; to the fact that the nature of man was afterwards to be assumed by the Second Person of the Trinity; to the delegated empire of this world which man was to hold. There are two expressions of St. Paul: that "man is the image and glory of God" (1 Cor. xi. 7), and that "the invisible things of Him from the creation of the world are clearly seen, being understood by the things that are made, even His eternal Power and Godhead" (Rom. i. 20), which seem to indicate that this record has a significance which as yet we can only partially understand. Then the story of man's creation is repeated in the second chapter, and while the other events recorded in the first chapter are very briefly summarized, that of man is very much amplified. This does riot necessarily indicate an independent account, as is sometimes asserted; at the fourth verse of the second chapter a distinct portion of revelation commences—the special dealing of God with man, and this could not be intelligible without an amount of detail with reference to man's origin, which would have been out of place in the short account of the origin of the world by which it is preceded. In this account the creation of Adam and Eve is recorded as two separate events, the latter of which is described in terms of deep

107

mystery, of which all that we can say is that they point to that still deeper mystery—the birth of the Bride— the Lamb's Wife from the pierced side of the Lamb. But in the case of Adam there is a remarkable difference from anything that has gone before. Two distinct acts of creation are recorded; one of which places man before us in his physical relation to the lower animals, while the other treats of him in his spiritual relation to his Maker. "The Lord God formed man (the Adam) dust from the ground (adamah), and breathed into his nostrils the breath of lives; and man became a soul of life. " The inspiration of the "breath of lives" distinguishes the creation of man from that of all other creatures.

The Geological records harmonize exactly with the Bible as to the date of man's appearance on the earth. It is towards the close of the age of gigantic mammalia, that the earliest remains of man's workmanship make their appearance in the shape of tools and weapons rudely fashioned from stone. Parts of human skeletons have also been occasionally found, but they are exceedingly rare. Weapons and bones are alike confined to superficial, and comparatively very recent formations. From such traces as have been found there is no reason to believe that any physical changes of importance have taken place in man's body since his first appearance on the earth. The differences which do exist are of the same kind as, and not greater than, the differences which exist between individuals at present.

The gift of dominion over the lower animals seems to indicate something different from that which gives one animal superiority over another, and accordingly we find that it is not by physical power that that dominion is exercised; but that in most of his physical faculties man is inferior to the very animals which he holds in subjection. It is partly in virtue of his intellectual superiority, and partly perhaps by means of an instinctive recognition on the part of the animals of man's higher nature (Gen. ix. 2) that that supremacy is maintained.

SECTION 10. DEVELOPMENT.

We have now to consider the question of development, in reference to the Mosaic Record of the last two days, and to the known facts to which that record has relation. The account of the third day's work has also a bearing on the subject, but as the same considerations will to a great extent apply to animals and to plants, it will not be necessary to make any special reference to it.

The facts in favour of the theory of development are these: −1. The different classes of plants and animals are not separated by broad lines of demarcation, but shade insensibly into each other. 2. The characteristics of the same species are not constant; the lion, for instance, the horse, the elephant, and the hyena of the present day differ in many minor points from the corresponding animals of the Tertiary period, so that unless there was a possibility of spontaneous change, we must assume successive creations of animals, with only trivial differences. 3. In all animals there are minute individual differences, and if under any circumstances these differences had a tendency to accumulate, they might in the course of time result in great structural modifications. 4. Man has been able to take advantage of this fact and by careful selection to mould the breeds of domestic animals to a certain extent in accordance with his own wishes.

The theory of development assumes that for the care of man other forces might be substituted, which in a long course of ages might result in changes of far greater extent than those produced by human agency. The forces assigned are natural selection and sexual selection. The difficulties in the way of this hypothesis have been already considered, and only require to be briefly re-stated.

1. As regards modifications of organs already existing, the two alleged causes are insufficient to account for the results which we witness, since in each individual case the concurrence of many contingent causes, continued through a long series of ages, is required to produce the result. But the probabilities against such, a concurrence in any one case are enormous, and against their concurrence in a large number of cases the chances are practically infinite.

2. That such causes do not at all account for cases in which an entirely new organ is developed, such as mammary glands—or for the case of man, in which intellectual superiority is accompanied by a loss of physical power.

3. That from the nature of the case it is impossible for us to ascertain that natural or sexual selection has ever acted to produce a single modification, however small, and that the results of man's

superintendence have not as yet passed beyond certain narrow limits, so that there is no justification for the assumption that such modifications are capable of being carried to an unlimited extent.

We see that in the only case in which change is known to have been brought about, it has been the result of choice and design. If then there is a probability that choice and design may have been exercised by a power higher than man, there is no longer any reason to doubt but that results much greater than any to which man can attain may have been brought about by the same means. And in fact the advocates of the theory of development do virtually admit the existence and action of such a power, whenever they have recourse to assumed "laws" to account for phenomena for which their naked theory can give no reason. For, as has been shown, law, if it is to be assigned as an efficient cause, and not merely as the statement of observed facts, can only be regarded as the expressed and enforced will of a higher power. And there was no reason why those minute variations themselves, which are the basis of Mr. Darwin's hypothesis, should be considered casual. Instead then of natural selection, or sexual selection, let us suppose that the selection took place under the superintending care of the Creator, and was directed towards the carrying out of His designs, and then we shall have no reason to doubt but that all results which consisted only in the modification of existing organs may have been obtained by the operation of those laws which we term natural, because they express modes of operation with which we are so familiar that we look upon them as automatic.

But there are other results for which no natural laws with which we are acquainted will thus account. Just as no mechanical laws within our knowledge will account for the rotation of the earth, so no physiological laws yet discovered will account for the changes when totally new orders of being came on the stage—when the course of life took, as it were, a new point of departure. But it is precisely at these points that the Mosaic Record points to a special interference on the part of the Creator. How that interference took place we are not informed. Very possibly it may have been the result of other laws which lie wholly out of the reach of our powers of observation. But whatever may have been its character, it does not in any way imply change or defect in the original plan, unless we know, (what we do not know, and cannot ascertain) that such interference formed no part of the original design. Everything bears the marks of progressive development, and there is nothing improbable, but rather the reverse, in the supposition that such a plan should include special steps of advance to be made when the preparation for them was completed.

The Mosaic Record tells us nothing about the method by which God created the different varieties of plants and animals. All that we read there is just as applicable to a process of evolution, as to any other method which we may be able to imagine. But it is remarkable that what Moses does say is just what is required to make Mr. Darwin's theory possible. So far then as the lower orders of creation are concerned, the hypothesis of development, modified by the admission of uniform superintendence and occasional special interferences on the part of the Creator, may be accepted as being the most satisfactory explanation that can be given, in the present state of physiological science, of the Scriptural Narrative.

But we have yet to consider this hypothesis as applied to man in Mr. Darwin's latest work. We naturally recoil from the thought that we have sprung from some lower race of animals—that we are only the descendants of some race of anthropoid apes. So long as it is asserted that we are no more than this, we may well be reluctant to admit the suggestion. But if it be admitted that to a physical nature formed like the bodies of the lower animals, a special spiritual gift may have been superadded, the difficulty vanishes. All Mr. Darwin's arguments with reference to physical resemblances may then be admitted, and we may allow that he has given a probable explanation of the method by which "the Lord God formed the Adam, dust from the ground" while we maintain that the intellectual and moral faculties of man are derived from a source which lies beyond the investigations of science.

The conclusions to be drawn from this investigation may be briefly summed up as follows: —

1. There is every reason to conclude that the process of Creation was carried on, in great part, under the operation of the system of natural laws which we still see acting in the world around us: such laws being so far as we are concerned only an expression of an observed uniformity in the action of that Being by whom the Universe was created and is upheld.

2. That inasmuch as the development of a new state of things differs from the maintenance of a condition already existing, the working of

these laws was necessarily from time to time supplemented by special interferences of the Creator, but that such interferences formed parts of the original design, and are not indications of anything in the shape of change or failure.

3. That many of the events recorded in the Mosaic Record are of the nature of such special interferences, while others point to remarkable developments of particular forms of organic life.

4. That these interferences thus recorded occur at the exact points at which natural laws, so far as science has yet been able to ascertain them, are inadequate to produce the phenomena which then took place, and that the developments are proved by geology to have taken place at the points indicated.

5. That the six days into which the work is divided by Moses do correspond to the probable order of development—that in three of them, the third, fifth, and sixth, this correspondence is marked by facts ascertained by Geology—that the fourth, in which no terrestrial phenomenon is recorded, corresponds to a very long period in the Geological record in which no indications of any new development are found—while the first and second indicate a state of things which the nebular hypothesis renders highly probable, but of which no positive information is within the reach of science.

Admitting then that there is something in the way in which the days are spoken of which we are at present unable to understand, we may yet confidently assert that such a record could not have been the product of man's thought at the period at which it was written. It is utterly impossible that it should have been the result of a series of fortunate conjectures without any foundation to rest upon, and scientific foundation there was none, for there is every reason to believe that the sciences which might perchance now supply some foundation are entirely the growth of the last three centuries. There is then only one conclusion that we can draw, that it is a revelation from the Creator Himself, and that if there is anything in it which seems inexplicable or erroneous, that appearance arises from our own ignorance of facts, and not from any error on the part of the Author.

Printed in the United Kingdom by
Lightning Source UK Ltd., Milton Keynes
136846UK00002B/244/P